DRYSTONE

A Note on the Author

Kristie De Garis is a writer, photographer and drystone waller. Based in Perthshire, Scotland, she lives with her two children and two ex-husbands. *Drystone* is her first book.

DRYSTONE
A Life Rebuilt

Kristie De Garis

Polygon

First published in 2025 by Polygon, an imprint of Birlinn Ltd.

Birlinn Ltd
West Newington House
10 Newington Road
Edinburgh
EH9 1QS

www.polygonbooks.co.uk

1

Copyright © Kristie De Garis, 2025

The right of Kristie De Garis to be identified as the author of this work has been asserted in accordance with the Copyright, Designs and Patents Act 1988.

All rights reserved.

ISBN 978 1 84697 646 9
eBook ISBN 978 1 78885 666 9

LOTTERY FUNDED

The publisher acknowledges support from the National Lottery through Creative Scotland towards the publication of this title.

British Library Cataloguing-in-Publication Data
A catalogue record for this book is available on request from the British Library.

Typeset by 3btype.com

Printed and bound in Great Britain by Clays Ltd, Elcograf S.p.A.

Contents

The Land	1
Stripping Out	35
Sorting the Stone	77
Laying the Foundations	115
Placing the Hearting	147
Setting the Throughs	179
Lifting the Copes	213
Afterword	221
Acknowledgements	223

For A and P

DRYSTONE

The Land

I'm alone. Too early even for dog walkers. Enjoying the solitude and a warm breeze, I walk towards the beach, passing remnants of planticrus sitting crooked in the field. Behind them, a couple of tired standing stones rest, leaning into their hips. The ebb and flow of a drystone wall catches my attention. Some parts still upright, some long fallen. I follow its undulating lines with my eye until the wall disappears thread-like into the distance.

No matter where you are in the world, north, south, east or west, if there is stone, there is drystone. Responsible for some of the oldest structures on our planet, drystone is a craft that's about doing what you can with what you have. In Scotland, it's everywhere. From the thousands of miles of field boundary walls, to the stoic gable ends of ruined croft houses, to remote stells keeping watch in quiet glens, the stones memorialise lives once lived.

Ahead of me, I see echoes of previous journeys etched in the marram grass. Beneath my feet, stone blocks litter the uneven ground, once part of the broch, now painted in lichen expressionism. Inside, snecks and jumpers lead my eye over the surface of faded, earth-toned walls: a pleasing, recursive chaos. Used as they were found or shaped with rudimentary tools; stones as big as a person, irregular and bulging, others smaller, stacked tightly and neatly like books on a shelf. Every one placed by hand more than two thousand years ago.

Moving to what remains of the steps, I rest my foot on the first. As I push my weight into the smoothly worn indentation at its centre, I imagine those who stood there before me. I climb higher and a gust of wind lifts my hair, and the dark strands dance a reel around my head while the waves move restlessly below. One hand on the stone, I raise my eyes to the horizon: that endless vista towards the Outer Hebrides and then a clear two thousand miles to Newfoundland.

I was eight years old when our mum moved us from one end of Scotland to the other. From Jedburgh, where you cross the border to England, to Thurso, where you catch the boat to Orkney. I don't know what Mum saw in Caithness, but I suspect it was just as far as she could go without leaving the country.

Caithness is flat and vast and lies prostrate to a wind pushed inland by the thuggish North Sea. A wind that reminds anything attempting to grow that it should perhaps reconsider. On the heaths and moors, plants have no choice but to stay close to the ground and each other. Even in July, when the rest of Scotland is dressed by the full flair of summer, the Flow Country is understated. Thriving in the nutrient-poor soil, bog asphodel, myrtle and bean hide among heather and peat. In the skies above these rare blanket bogs, golden plover and greenshank float silently on warm air and mottled wings. On the coast, the sandstone shark fins of Duncansby Stacks ward off all but the most determined travellers. The Borders are Scotland at its most pastoral, but Caithness, akin to a moonscape.

On either side of the road, cutting long and straight through an expansive moorland, people stood in smoke and flames, soft-edged silhouettes as they beat the ground around them with huge paddles. Muirburn, the controlled burning of heather. It felt ancient, apocalyptic.

Some parts of Scotland coddle you with rolling glens cradling picturesque hamlets, and views that are broken by the shelter of forests. In Caithness, there is no respite. It's onwards over difficult ground, or nothing.

•

I'd spent the last few years slowly piecing together that we were poor. While my brother Matthew and I would eat dinner at the table, Mum would work around us in our small kitchen. I had assumed she was too busy to eat, that she would serve herself later;

but one day, sitting in the living room, kitchen door ajar, I saw her scrounging leftovers from our plates. When the school sent a letter requesting that we bring an apron to cover our clothes for art class, my mum announced, dragging out the sewing machine and cutting up one of her old dresses, that we would not be buying anything. At school the next week, I watched as our teacher tied bows behind twenty plain pinnies in bold, block colours while I struggled with blouson sleeves and billows of floral nylon. Yes, it was the eighties.

I'd seen how kids from Jedburgh's rich families were automatically popular in school, how the staff spoke of their parents with respect. I tried to ignore the comments about my father's absence and how my mum was trying to steal people's husbands. At Halloween we made sure to stick to streets that looked like ours. We didn't dare venture into the cul-de-sacs where people had red-brick bungalows, gardens so sprawling they could accommodate full-grown trees, separate garages with brand-new Volvos inside.

So, pulling up to our new home that rainy Tuesday, I was impressed. Four large triangular sections of gable end protruded from a vast white frontage, and the main door was wide as a car. A mansion! It had its own glass-panelled foyer and its tall corridors were bordered by wooden doors. Like an expensive hotel from a film. Even the name sounded like it was from a story. Fairview.

Maybe in Caithness we were rich.

•

Nervous, I calmed myself by imagining that my new classmates at Halkirk Primary would be impressed when they found out I lived at Fairview. Standing in the playground, I smiled tentatively at the children willing to make eye contact, until a girl approached me.

'Hi. I'm Anita. What's your name?'

Before I could answer, I heard a voice say, 'Anita! Why you talking to her?'

The voice belonged to a stocky kid whose sandy-coloured hair stood on end like the popped collar on his school shirt. Appearing from within a crowd of boys, his influence was clear when Anita turned away from me and rejoined the huddle of girls by the wall.

Despite her contrition, the boy continued, shouting across the playground, 'She's a tink, Anita. She lives in the poorhouse.'

At the time I didn't know that Fairview Flats had once been known as Thurso Combination Poorhouse. Nor did I know that even after it had changed use and over a hundred years had passed, the building couldn't shake its reputation. Our fairytale mansion had been a charity shack. Small towns have trouble forgetting.

When I got home from school that afternoon, I wasn't sure I was going to tell my mum. I didn't want her to feel the same humiliation that I did. Walking through our living room as quickly as I could, I dropped my rucksack by the three-bar heater that was glowing orange and went straight into our closet of a kitchen. Busying myself with making toast, I heard my mum's voice. She sounded tightly wound. '"How are you, Mum?" . . . Oh, yes, fine, Kristie, thanks for asking. How was your day?'

Between her annoyance and hearing the words 'How are you?', the tears came. I picked up my slice of barely toasted white bread, began to eat and tried to swallow my sobs.

Appearing at the kitchen door, my mum smoothed her tone before asking, 'Och, what's wrong?'

As my chest started to heave, eating became impossible.

Mum came closer, and with the same care she would use diffusing a bomb, removed the wilting piece of toast from my margarine-smeared fingers and placed it on the crumb-covered counter. She put her arm around me and pulled my head towards her body. Slowly I relinquished.

With a pat on my shoulder, she took a small step away and addressed me in her I-mean-business voice, 'So what's going on?'

The words tumbled out. 'Gary said I was a tink and told her not to talk to me and they said this is a poor house and that we're poor and that I have nits and no one should play with me and . . .'

As I took a breath to check my mum's reaction, I saw the shadows of guilt moving across her face. Even through my blurry eyes I could tell that she already knew about 'The Poor House'.

I began to cry again. I felt her arm on mine and resisted a little, before eventually allowing myself that hug. But I felt betrayed. Not just because she had moved us here, but because she had allowed me to believe, for even a second, that it was something to be proud of.

It was only after I had learned our home's true identity that I admitted to myself that there had always been shouting. The raised voices of drunk men in the hall, Mum heating milk to distract us with hot chocolate. Sitting in silence, we would watch the door until the banging had stopped and the shape of a stranger's feet had moved away from the gap at the bottom. One night, Mum told us to stay away from the window, to ignore the sounds of a fight. The boots sliding in the gravel, the grunts and the swearing. At school the next day, all anyone could talk about was the stabbing at Fairview Flats.

•

Mum had used the move across the country to change our family name. She dropped Mohammed for the name of my father, a man she was never married to. He had failed us, but now his name could make our lives easier. Unless someone asked, we didn't need to mention that we were 'part Pakistani'. With amber eyes, dark hair and light skin, I didn't really stand out too much among the other kids. I probably could have avoided trouble if I had kept my mouth shut. I'm not sure why I didn't.

My heritage became another weapon in the bullies' armoury. Not only was I poor, a 'tink', but now I was a 'Paki' too.

I could see the commotion starting in the far corner of my vision. The boys jumping around, their forms growing wider and closer to the ground. I turned my body away, pulling at the hem of my shorts. When I heard the animal noises 'Ooo! Ooo!' and laughter, I walked to another exercise station. Waiting my turn to dribble the ball, I looked at the coloured cones on the floor, and pretended everything was fine.

'Oi, Monkey Girl.'

I turned my head.

Immediately, I understood that this was a grave error. Gary looked excited, thrilled even, as I fought hard to push back my tears. Turning to his friends, he shouted, 'See, she answers to her name!' before basking, bright-eyed, in their approval. Then he faced me again, put his hand up, set an affected, snooty look on his face and said, 'Go back to the jungle where you came from, you hairy Paki.'

I couldn't tell my mum this time. I'd seen how her eyes had grown more anxious, shoulders higher, and I'd heard her crying behind the closed door of her bedroom. Thinking I was being clever, I communicated my worries in question form.

'Mum,' I asked, watching her fold a blanket onto the edge of my bed, 'did anyone ever call you names at school?'

Looking right at me, she ignored my attempt at subterfuge. 'Kristie, never, *ever* let them see you cry. If they don't know it hurts you, they'll get bored and move on.'

I took her advice to heart, pushing away fear with anger, and the anger away with my fear of showing it. I never, ever let them see me cry. And, one night, I rid myself of the source of my problems.

We were forbidden to touch our mum's razor. I sneaked it into the bath with me, hiding it in water made opaque by Pears soap and Mermaid Matey bubble bath. Trembling, listening for Mum's footsteps, I ran the blade along the soft, dark hairs on my thigh. Back then, razors were a simple affair, no lubricating strip, no ergonomic design. It was a single blade, crudely housed in cheap plastic.

I dropped it at the first cut. Looking towards the pain, I saw a slash of red on my pale skin. I'd had no idea that you could cut yourself! I watched as the blood slid down my calf and into the bath, where it remained for a split second before dispersing. I knew I should stop, but the skin before the cut was smooth and hairless like the girls in school. The thought of another PE lesson was enough to spur me on. I cut myself again, and again, then got ready for bed, stinging, but relieved.

Sadly, the water did not wash away my sins. When she cleaned the bath before settling us, my mum found hundreds of tiny black hairs lining the white ceramic tub. Appearing at our bedroom door, razor in her hand, she thrust it towards me, shouting. In panic I retreated to my bed, pulling the covers up around my ears. As she tried to pull the blanket away I fought against her, wrapping the wool up in my fists.

But she had that mum-rage, and it took only a few hard tugs to wrench the covers from my hands and body. Forcefully pulling up the legs of my pyjamas, she saw the cuts and the smeared, dried blood. With her right hand raised close to my face, she told me never to do it again or she would send me away. I could smell her Oil of Ulay, Silvikrin hairspray and cigarettes as I nodded.

•

I'd waited, as usual, to be the last one in the changing rooms. When I rolled down my tights, I saw that the hairs had grown back enough to be noticeable. Today was dodgeball and I didn't want to stand in front of everyone. I didn't have it in me to remain aloof while some shouted insults and others listened. And so I pulled my tights back up and put my shorts over the top before realising that this would also draw attention.

Standing on a bench, I reached for a small window in the wall above me. It was stiff, and I struggled to push it open while balancing on loose wooden slats. Sliding against the painted bricks of the wall, I mustered the strength to pull myself up to the windowsill. Arms shaking, I wiggled through the tight space and paused for just a second before jumping.

As soon as my off-brand trainers touched the grass, the pounding in my chest almost knocked me off balance. I had never broken a rule in my life, and I knew that Miss Swanson would soon come back into the changing rooms to see what was taking me so long. So I ran. It's a seven-minute walk from Halkirk Primary to Fairview Flats but on that day it took me one hundred hours.

My lungs burned, eyes stung and streamed, and my legs resisted, wobbling, threatening to give out at any minute. I reached the soggy field that stretched from the edge of the playground to the road I needed to cross to make it home. It would take too long to go around it, so right down the middle I ran. With dirt spattering my calves, I kept turning to look behind me, to see if anyone had followed.

At last, I stood at the main entrance of Fairview Flats. Buzzing again and again, I grew even more frantic. Crying, I held down buzzers until someone finally let me into the building. Inside, I pressed my head against our door, knocking and calling to my mum until she answered. Her expression when she saw me, dishevelled and soaked in mud, arms scraped from climbing through the window, was unreadable.

She ushered me into the flat and closed the door, locking it at the snib and then with the keys. Finally turning to face me, she put her hands on her hips. 'What . . . what are you doing here?'

With the adrenaline wearing off, I had no idea what my story was.

'Hang on. Where's your coat?' She looked angry. 'Where's your bag?' Suspicious.

I'd already lost one coat, ripped another, and my mum had warned me I wouldn't get another. 'They're at school, they're all fine. I promise!'

I'd expected her to be happier, but her tone remained firm. 'How did you get here? What's going on?'

I could think of only one thing to say. 'PE.'

Her voice rose. 'PE?!'

'I have to wear shorts,' I added.

Shaking her head, she continued, 'I'm sorry, Kristie, not liking something doesn't mean you can run out of school whenever you want.' More head shaking, as she moved towards the coat hooks near the door. My eyes widened as she picked up her jacket and fished around in one of the pockets. Extracting the car keys, she gestured. 'Come on, then.'

It hadn't crossed my mind that I would have to return to school, never mind that my own mother would deliver me back into the clutches of the bullies. But what choice did she have?

My mum was not even thirty and navigating life as a single parent, carrying the weight of her trauma, living in entirely white and often deeply racist communities, in poverty and with no support network. She left our father when she recognised what life would be with him in it. When she understood what our lives would look like in the Scottish Borders, she left that behind as well. It was a local butcher who asked her, 'How did a decent woman like you end up living in a place like that?' Soon, she would leave Halkirk behind too.

•

In the way that children do, when I was told that my grandfather Faqir had left Pakistan 'because of The Partition', I accepted this explanation without understanding. It took until my late twenties to learn that the Partition of India was one of the largest forced displacements of humans in history. 'Because of The Partition', which sounds like someone accidentally put up a fence in the wrong place, was simply a way to stop a child from asking questions, to protect them from the answers.

Soon after arriving in Scotland, Faqir met Catherine at a party in Glasgow. In the 1950s they were a couple who stood out. Him, a six-foot-plus Pakistani man with a long face, hooded dark eyes, brown skin, black hair and her, a small, ruddy-cheeked, blonde-haired and blue-eyed Norwegian-Irish woman. Six children quickly followed. Three girls and three boys, my mum the youngest of them all.

Seeing us eye his arm resting on the back of Nana's floral sofa, Mum had pulled us aside and quietly explained that the bumpy scar sticking out from under Uncle Zahid's rolled-up sleeve had been made by a sword. Our young imaginations conjured up swashbuckling pirates and brave knights. Swords belonged to those worlds, not to ours.

Mum relayed the details of her childhood in the same way someone might tell you about their day at work. Between sips of tea and cigarettes, she said, 'I turned the corner and a man was pointing a gun at my brother's head.' Arms around us on the sofa, her face filled with pride, she told us, 'And Nana said, "That's my son," and punched the gang leader in the face and he fell backwards over the fence!' We were thoroughly entertained. 'Tell us again!' And she would.

My family weathered so much, and these collective injuries bound them until the death of my grandfather, three years before I was born. That loss tore them apart. I felt their love, but also the sharpness of their pain, which, unspoken, persisted as the broken edge of my family. Fallings-out and estrangements were a regular occurrence, and days, sometimes weeks or months, would pass before we saw our relatives again. Any comfort I found in family was accompanied by a knotted stomach.

•

It was a short walk from where we lived in Jedburgh to the home that Nana shared with our aunty. Waiting for us in the front garden, Nana would be wearing a tabard, like our dinner ladies wore at school, the bright red birthmarks on both her arms showing like faded tattoos. Our aunty would be busy in the back garden, pinning up peapods, or in the living room, knitting yet another itchy jumper, watching TV.

We only ever used the door at the side of the house. Shoes off in the porch and into the tiny square kitchen that smelled of curry, broth, mince, cigarettes, or a combination of a couple of those things. Always a pan loaf on the go, endless cups of tea and a biscuit tin that couldn't shake the smell of Mint Clubs, even if the last time it held one was a year ago. Even in meagre times you were guaranteed to find a squashed Wagon Wheel or broken party rings rattling around in its copper-coloured bottom.

Once biscuits were selected, we were sent to the living room. Not really a room for children at all. Brass ornaments that still smelled of Brasso, cottage-shaped teapots and porcelain figures kept behind tall glass doors. Knitting patterns like complicated maths problems lay next to balls of wool bigger than my brother's head.

When the fire was on, we would sit on the floor pushing our faces so close to the flames that I wondered if the stinging I felt in my cheeks was a sign of permanent damage. Sometimes we took turns prodding the coals with an iron poker, one ear on the door. My aunt had warned us about the white-hot sparks that could fly from the hearth. Pointing to little holes burned into the carpet as evidence, she always put the mesh fireguard in place and, before leaving us alone, turned, looked into our eyes and said, 'Don't.' Ascertaining the whereabouts of my mum, nana or aunty was easy because all three of them wore, from wrist to mid-forearm, identical silver bangles from Pakistan. And all of them were fond of gesturing dramatically.

Once we grew bored of the fire, or the heat made us sleepy, we would tip out the wicker basket of VHS tapes. Some films had been bought brand new and came in hard cases with the movie poster framed on the front. Others were in flimsy cardboard sleeves with multiple handwritten titles scored through. Taped from the TV. My favourite was *Little Women*. I loved the dresses and the close-knit family, but I always fast-forwarded the part where Beth died. My brother's favourite was *Moonwalker*, and he would rarely sit still when it was on, compelled by his own internal logic to stand in the living room copying all of Michael Jackson's dance moves in real time.

If it happened once, it happened a thousand times. Shouting would waft from the kitchen alongside Radio Scotland and the smell of what was cooking. And it would grow louder until it became the audio track to whatever we were watching; Michael Douglas's mouth would move, but it was an angry Scottish woman's voice coming out.

Looking at my brother, I'd sit up, fingers on the ground like a sprinter, ready for the signal to move. I'd raise my eyebrows, a warning, but he'd respond with an annoyed, exaggerated shrug and stay where he was, lying on his stomach.

When he heard her bangles jingling, he'd fly up off the floor like he'd been scorched by one of the embers my aunt had warned us about. And then Mum would storm into the room. She'd grab our coats and jumpers and tell us, 'Get up. We're leaving.'

I remember once, being on my knees and trying to put the VHS tapes back into their basket when I felt my mum's hand hook into my armpit, and close to my ear, 'I said now.' Her voice a low growl as she pulled me to my feet.

Pushing Matthew in front of her and dragging me behind, she directed us towards the kitchen. Nana and Aunty stood on the other side of the breakfast bar, arms folded, as we were hurried into the small porch. Struggling with our shoes as the argument continued, my mum shouted responses over her shoulder, 'Oh aye, it's aaaaaalllllways me, isn't it?'

We were shoved through the door, laces loose, then wrangled, briefly, back into the kitchen and sent towards our nana and aunty with firm instructions to say goodbye. We understood the implication in our mum's tone. She meant 'for ever'. Dutifully, we completed the task, unable to hold back our tears.

My aunt, disgusted, said, 'Look how much you're upsetting them, Allison.'

Mum's face exploded into incredulity. 'Look how much *I'm* upsetting them? Oh, you've got a cheek. You're the one who'd rather not see us at Christmas. Did you know that, kids? Your *precious* nana and aunty don't want to have you at Christmas!'

Passing us back and forth like props, each side tried to prove their innocence through the hurt the other was causing.

My brother's small body drew tight next to me. I looked down and saw his hands balled into fists. Arms rigid, face red, he burst forth with a scream: 'STOP IT.'

For a few seconds, everyone stared at Matthew. In the silence, I wondered if he had somehow negotiated a ceasefire.

Then, voices shouting all at once.

'See what you've done!'

'I'm done with this!'

'We're leaving!'

Walking home, we struggled to keep pace with our mum. By the corner of Brewster Place, she was ready to talk, and by the lane between the garages and the flats, she was in full flow. How she had been pushed out, how unfairly she was treated, how our family didn't give a fuck about us. How we only had each other.

•

Robert looked like a 1990s red-haired Glaswegian Steve Zissou, and indeed his self-built catamaran sat at a lonely bend on the River Clyde. The boat was half finished, and exposed fibreglass pressed rough cross hatchings into our hands as we crawled through its confined spaces. Robert told us that when the boat was ready, he would sail us around the world. We would lie on a net slung between the hulls as we skimmed over turquoise waves. We would anchor near tropical islands, eat the fish we caught for dinner, and sleep out under the stars. Or we could do 'the next best thing' which was, of course, a jaunt to Northern Ireland.

I'd never been 'abroad' before, so the long day of travelling from Thurso to Belfast was exciting. Robert met us at the ferry port and we followed him in our car to a house in the countryside. It was big and shabby and sparsely furnished. The blankets smelled like old lady and damp, but, cold in bed that night, I tucked them around my neck.

The first day of our trip, Robert drove us through Belfast. We gawked at the landscape of a war zone. Bold murals with indecipherable themes painted on gable ends. I wondered if the bullet holes were painted on too.

From my place on the back seat I heard my mum whisper, 'I'm not sure about this.'

Robert cleared his throat but didn't say anything.

'It's important,' he said after a long while, 'for kids to see real life.'

At some point, I fell asleep. When I woke, I looked over at Matthew, eyes closed, clutching his SuperTed with the tiny, red homemade cape. Looking out of the window, I saw it was growing darker, and the tall buildings of the city had been replaced with dingy streets of pebbledash, gardens sprouting mattresses and rusting white goods. On the outer walls of homes and shops, paintings of crossed flags, defiant faces and men holding their guns high. The few people we saw stared at our car, following its journey for longer than they needed to. My legs were squashed against the backpack that was jammed down by my feet. I tried to stretch them but found myself caught in place, my trainer tangled in one of the straps.

We stopped abruptly and a grey truck pulled up beside us. It looked like one of my brother's army toys. Two uniformed men, guns slung over their shoulders, got out and started walking towards us. My mum looked at Robert. He said, 'Let me do the talking.' Matthew woke up as the men rapped on our car windows, gesturing for the adults to get out.

My mum looked back at us as she opened the car door. 'Don't worry,' she whispered, but I was already crying.

We watched them. Robert was animated, laughing, while the men's faces gave nothing away. My mum stood silently, as she had been told to. It was taking a really long time.

Matthew, his panic reaching a crescendo, unclipped his seat belt and said, 'I'm getting out.'

I pleaded with him. 'No, Matt. No. Please don't.'

As his door opened, the heads of the men with the guns snapped towards the car. One of them shouted. Matthew froze. Mum said something quickly to the man and before he could reply, she was running towards us, saying, 'No. NO. Stay in the car!' before slamming the door and returning to her place beside Robert.

We waited, growing colder in the fading light, watching helplessly through windows slowly made misty by our breath.

Eventually, one of the men, his face impassive, walked back over to the truck, rummaged in the cab, and returned with a map. Waving Robert closer, he traced his finger over the paper, stabbing a certain point several times, saying something. Robert nodded emphatically.

Then it was over. The men returned to their truck, and Mum and Robert got back into our car. Both facing forward, they said nothing as Robert turned the vehicle around and drove us back the way we came. Matthew and I looked out of the rear window, watching the truck follow us for a while before it turned down another street. Then streetlights became more infrequent, and soon they were gone, but still no one said anything. The tension lasted through dinner and was only broken when, from my bed, I heard Mum and Robert shouting.

The next day, we were already on our way to a country park when Mum announced confidently that she wanted to cut the trip short and go home.

Suddenly, a chorus of horns blaring around us. Everything was happening in slow motion and I couldn't really hear any more. I saw my mum, tears pouring down her face, pleading, then angry, but Robert kept driving into the oncoming traffic.

When my mum screamed, 'You're going to kill us!', everything sped back up. Without thinking, I pulled off my trainer and held it in my hand. Scuffed-up white, with an energetic teal shape on its side, the laces were tied in at least eight double knots. I lifted it up, reached forward and hit Robert as hard as I could across the side of his head.

Eyes on the road, grabbing the air behind him with his free hand, he tried to take the shoe from me, but I was too fast. I raised it and hit him over and over. He thrust his arm into the back of the car, reaching for me and sneering, 'Are you going to fucking cry?'

His ear was red where my trainer had made contact.

The sound that came out of me was guttural. 'I will never cry for you!' I screamed the sentence as loudly and as long as I could, my voice eventually breaking in my throat.

And then the world was silent. A brief moment of solace before it all started moving again and the sound of horns and my mother's shouting returned. I have no memory of the trip after that point. I don't recall what Robert said as he drove us back to the house, whether I put my trainer back on, or anything about our journey home to Scotland. But I do know that we never spoke about what happened. No debrief. No play by play. No 'That was fucking wild, right?' No 'Are you OK?' Nothing.

A few weeks later, Robert turned up at our door with a puppy. We named him Moss and spent the day pulling him around the living room, squashed into my brother's John Deere trailer, while Mum and Robert fought in her bedroom.

•

To be one of the hundreds of McKays, Gunns or Sutherlands in Caithness meant belonging. The longer your ancestors had been there, the better. Some people could trace hundreds of years of their family tree to one village and were very proud of it. I was proud of my family too, of how we were from everywhere – many villages. And as I got older, I began to more openly embrace my Pakistani heritage, crowing about it to anyone who would listen and many who would not.

I would abruptly drop it into conversation, convincing myself that whatever tenuous thread I had followed was justification enough. Talking about your aunty going on holiday to Greece? I was going to respond by telling you that 'my granddad was from Pakistan'. Putting it out there felt like ripping off a plaster. If someone knew from the start, then I didn't need to worry that they'd ever find out. I was wounded and I was angry. I walked a fine line between what I felt and what I would show.

One morning, walking up to the hatch of the tuck shop, I saw a group of sixth-year boys who regularly tormented me. The dark, fine hairs on the back of my neck stood up. I readied myself.

There was a rhyme that I heard a lot in school. I don't know if someone bestowed upon me the peculiar honour of having composed it specially, or if other kids across Scotland were subjected to it.

'Hey, Pakistani, does your granny have a fanny?

Does it smell?
Fucking hell!'

The boys at the tuck shop began chanting this masterpiece of racism and misogyny, and as they approached me, one of them bent down and sniffed loudly between my legs.

'It does smell . . .' He stood up triumphantly. 'Of curry!'

I died inside, but my anger took over, soothing me in a way I had learned to rely upon. 'Fuck off,' I said, in as casual a tone as I could muster.

They didn't fuck off. Instead, they pelted me with pocket change at point-blank range. I adopted a defensive stance, my eyes cast down, while one hand protected my face. I listened to them shout 'Paki' as the coins fell around my feet like spent ammunition shells.

Puzzled by my outward insouciance or, more likely, having run out of low-denomination currency, they eventually stopped and began to drift away. Waiting until their backs were turned, their attention on each other instead of me, I bent down. Hands shaking, I tried, as quickly as I could, to pick up the copper pennies lying flat on the tarmac.

With their money tight in my fist, I walked to the tuck shop counter and used it to buy snacks, carefully counting it out to the shopkeeper, who was desperate to avoid my gaze. Heading back towards the main building, I felt sick as I tried to eat Space Raiders in a way that would convey victory, or at least indifference. I made it to the dark rooms in the art department before I began to cry.

•

We followed Mum into the dim hallway and listened as she ordered us in a tense whisper to stay there and not move an inch. Someone was on the roof. And she pointed silently towards the scrabbling and banging coming from above us. Matthew, eyes sleepy, was keen to not be living this nightmare. 'It could be a fox or something,' he said quietly, as Moss, safe in his arms, lifted one small ear.

Her own eyes on the ceiling, Mum reassured us, 'It's just . . . someone playing a joke.' I wasn't sure I believed her. We looked nowhere but up, tracking the heavy sounds. Our mother, poised, ready to defend us, or kill us if we moved.

Someone chapped the door. Hard. I jumped. The knocking came again and Mum put her finger to her lips. She prised Matthew's hand off her nightgown and walked into my bedroom to peer behind the curtains. She came back, swiftly.

As she approached the front door she seemed to remember us standing there and turned her head, face profiled in porch light, to tell us to go back to our rooms. We did, but I stood with my left ear against my bedroom door, barely breathing.

The police officer greeted her.

'Did you get him?' she said. 'He was on the roof. I saw him in the skylight.'

'Yes. He was still there when we got here, wearing that, uh, ski mask–'

'Do you know what he was doing up there?'

'Don't worry, he won't be doing it again. He's in the van. We've made it clear he has to leave you alone.'

Her voice was trembling. 'You tell him, you tell Robert, that if he ever touches my kids, I'll wring his neck.'

•

Caithness had no cinema, barely any public transport, and no shopping centre – just a Woolworths with a CD section and a pic 'n' mix that Mum assured us was extortionate. The internet was a faraway fantasy, and Mum wouldn't let us have a TV, often expressing her concern that shows like *Home and Away* or *Neighbours* would rot our brains. But I suspected it was because we couldn't afford a TV licence. And for a lot of my childhood, I didn't have firm friends. Even for the nicer kids, it made sense to distance themselves from a person who attracted regular hassle. So I spent a lot of time with my brother, playing outdoors.

The moorlands that surrounded our home required skill to navigate. My brother and I learned this the hard way, often traipsing home wet to the knee, one sock stained bronze by peaty water when we had foolishly taken the solid look of mossy ground at its word. In the summer we discovered tiny life hidden in the grass, the drama of a slow worm sighting, the smell of sphagnum. In winter our playtime was scored by the music of ice in farm-road potholes stretching and creaking beneath our feet. Braving a white-out in pyjamas, our hats, scarves and coats thrown hastily on top. Returning home to the smell of burning peat, our bodies raw with cold.

Our play was dictated by the seasons. In winter, quickly fading light set a strict curfew, but in summer we could take our sweet time. Around solstice the light lingered right above the horizon all the way through to dawn, and with such a gentle transition between day and night, an appropriate home time was much harder to determine. It took just a few bike rides through moorland in light that showed you too much (but not enough) to convince us of the benefits of better timekeeping.

Summer was a whole season of carefree miscalculations. No jumper or jacket for the cycle home, shorts and T-shirt feeling brutally inadequate as the sea air, pulling in the evening mist, covered our arms and legs in waves of goosebumps. In the morning, an eight-mile cycle had seemed easy, so we had given no thought to saving energy over the course of the day for our return trip. One eye on the shifting shadows of the land, we would pedal hard and fast, until we closed the gate between the moor and the track to the farm cottages. From there, we would push our bikes so we could lean on the handlebars for support, arriving at our front door, exhausted and close to tears.

As soon as we were inside, our mum would empty our stuffed pockets. The animal bones were immediately confiscated, the flowers we picked for her were unbent, fluffed and put in a vase, and our collections of stones were piled onto windowsills all over the house. Any parent knows that a walk with kids means hauling five kilos of stones back to base. Children know what we pretend to forget: stones are treasures. Even as adults, we will privately pocket a pretty pebble on a beach, and don't we all secretly hope to turn one over and discover the imprint of million-year-old life? Within their small, solid forms, stones encapsulate the very essence of memories.

Frank, another of my mum's boyfriends, looked hastily sketched. No solid edges, each part of him an impression of the real thing. His long, pony-tailed hair and pointed beard reminded me of a Buffalo Bill I had once seen in a performance of *Annie Get Your Gun* at the Playhouse in Edinburgh. As if in homage, Frank wore a suede jacket and tall leather boots.

He took us on a walk in the field next to his house, asking us if we knew what had made this track or left that dropping. We did not, and we looked on horrified as he picked up a small pellet of shit from the ground and tasted it. Not popping the whole thing in his mouth but just kind of licking it. 'Aye, that'll be a deer,' he said before walking on ahead.

But I loved the floors in his house.

Standing lonely somewhere outside Halkirk, Frank's house was a tiny, loaf-like structure that sat so low in the moorland it looked like it might be sinking. Huddled within the metre-thick, solid stone walls were a few deep-set windows that were visible only at night, when interior light illuminated their positions in the dark. I don't remember doors; instead, heavy, faded curtains hung on sagging rails. Wires snaked the walls, secured occasionally with white tape curling and blackened at its edges.

Laid in huge squares, flagstone floors ran throughout the building. Ripples, the influence of water frozen in time, disrupted their surface, making them look like a nighttime riverbed. Ice-cold in winter, the slabs warmed as the ground did, and in summer you could walk on it without socks, feeling every lithic detail beneath your feet. The stone was always warmest around the hearth where peat, cut in blocks from the land, fuelled a fire, the sole source of heat in Frank's home. A permanent draught from the front door excited ash in the grate and sent dancing particles through the air to settle all over the house.

Frank, although not particularly houseproud, seemed to think the flagstone warranted special treatment. First, he'd sweep, paying particular attention to the indentations between the stones. Then, fetching a bottle of milk from the fridge, he'd fill a small bowl with the cold liquid and carefully set it down beside him. Soaking the corner of an old red rag, he'd gently wave this cloth across the floor – and the flags, at first dusty and dull, would reappear from beneath the cloth, an oil slick of dark, shining stone.

Like Caithness itself, Caithness flagstone is very flat and very tough. An obstinate, sedimentary rock that splits along its bedding plane to create sheets, it reflects colour the way the sea does. Reacting to the moods of the sky with intense clarity, or murkiness, over the course of a few hours it will shift from uncorrupted black to leaden grey to warm-toned, to blue.

I saw natural deposits of flagstone in coastal areas or in long-abandoned inland quarries. On the moorland itself, I would find it in the form of an existing structure – or a structure that once existed. Scattered across Caithness is thousands of years' worth of drystone: brochs, cairns, blackhouses, stells, fanks and, the very bones of the land, endless miles of drystone walls.

Drystone is a traditional craft. Building with stone without mortar, you can create something as simple as a wall or as tricky as a chapel. Drystone is about using what you have to do what you need. What Caithness has is flagstone.

Rising like a spine from the ground, drystone is often the only visible feature in the recumbent Caithnessian landscapes.

A constant roadside companion, tightly packed stone walls line ditches amid the white froth of meadowsweet and the always surprising burgundy and apricot of water avens. Running the length of fields, flagstone fences – those single, square sheets of flagstone placed edge to edge – stand like domino-ed gravestones. A memorial to the land and all that has been lost to time.

From an early age, my hands knew the weight of stone. Firmly pulling a weathered grey sheet from a line of copes was like unsheathing a heavy blade. Misjudging the integrity of a wall, it would shed into a slithering clatter. On a sunny day the smell was of dry moss-musk, but when I lifted a well-settled stone to reveal a damp nook, it would be stirring with quiet life and metallic earthiness.

One afternoon at school, we stood in line to peer at a slide containing the inner epidermis of an onion bulb that had been slotted into a microscope and placed at the front of the class. I waited, bored, to take a look at what would most certainly be a bunch of boring onion shapes. When my turn finally came, I pressed my eye against the warmed black rubber of the eyepiece and saw, unmistakably, the interlocking shapes of a drystone wall.

I began to pull stones from the ground like root vegetables, brushing off the dirt and stacking them in an attempt to replicate that neat cellular structure. As I built these simple altars, I'd collect fronds and flowers from the banks of the burn and place them on the stones. A wee pagan, arranging pink mops of ragged robin among green fern fingers, I hoped the universe would be pleased with my small offerings.

•

I am older than Matthew by eighteen months but I deferred to him on most counts. Even from a young age he seemed so capable and sure of himself. While I was still performing a risk assessment, he was up and over the high fence, never worrying about barbed wire, or if we should be climbing at all. My brother could persuade me to do something I really didn't think I should be doing. Sometimes he would be gentle, all sweetness and 'Come on, Krist, it'll be fine'. At other times he was ruthless. I felt torn between being flattered that my brother wanted my company and worried that he was using me as the fall guy.

Matthew had our mum wrapped around his little finger. I felt bound to tell the truth, exactly as I saw it, and marvelled at his ability to be so considered. He always knew what to say. I was always saying too much.

This translated to life outside our home, too. At school, Matthew was successful in forming friendships, seemingly able to understand what it was that people expected of him. He aggressively enacted a life strategy of aggressively fitting in.

He wanted brand-name shoes. I once asked the woodwork teacher to stick my shitty trainers back together with his glue gun. He joined the rugby team and I joined the anti-bullying committee (with the rest of the kids who were being bullied). Perhaps he so enthusiastically pursued acceptance because for him it was attainable. Even if it meant leaving me behind.

Our peers played along. They knew he was my brother but didn't really hold him to it. He didn't really hold himself to it either. Matthew not only kept quiet when others were calling me 'Paki'; he did, on occasion, join in. And everyone witnessed the power and decisiveness of his words, none more than me. I saw my brother as influential, not just in the sense that he had the support of our peers, but in the sense that he didn't seem to hold anything particularly sacred.

My brother's middle name is Faqir. Each member of our close family has a Pakistani name: my uncles are Zahid, Anwar and Sajjad, my mum's middle name is Fatima, and my aunts, despite their very British first names, have Mohammed as their surname. I am the only one who doesn't have a Pakistani name, but, back then, I felt sure that I had earned one. Certainly, more than my brother.

I wish I had known then that he acted the way he did precisely because he understood what our family had been through. Most of the stories we had been told about racism involved a male family member being seriously hurt. To Matthew, my proud proclamations of being 'part Pakistani' were both a loose-lipped sabotage of his social status and an explicit threat to his safety. I see now that in our fight to survive, neither of us stopped to consider the other.

•

Early starts in alcohol were common in Caithness. Teenagers would gather in the run-down parks of run-down towns, passing round brown paper-bagged bottles under the tungsten glow of streetlights. My mum noticed too, and I suppose there was some logic in wanting her children to experience alcohol within the safety of their own home.

She would buy stubbies from Safeway for us to share with friends. Cheap to buy, these short, fat bottles of weak beer tasted more like the smell of something. In their little cardboard cartons, they looked like green bowling pins carefully racked before play. Post-game, they would be strewn across our bedroom floors alongside empty crisp packets and CDs.

The older girls, friends only on the bus and outside school hours, had no interest in drinking beer with my mum. They assured me that the pub's door staff didn't care if you were underage, even fifteen, as long as you were pretty.

Ignoring my stomach's spirited somersaults, I handed the doorman my obviously fake ID. Created at home on our PC, brought to life by our cheap printer, and laminated during a quiet lunchtime in the art department, it was surely my ticket to an automatic rejection.

The girls had gone on ahead, not wanting to risk refusal by association. Eyebrows pencil-thin, straight hair parted in the middle, knee-high boots, paisley-patterned camis over denim skirts, they'd had no problem getting in. I stood, cold, feet already sore in a pair of borrowed heels, as I waited for the doorman to decide whether I was pretty enough to break the law.

To think we handed so much of ourselves to some forty-year-old divorced guy called Dave who wasn't allowed to see his kids.

Inside, my eyes travelled from the pool table to the tangled speaker wires, to the surly bar staff. This was not what I had expected. I saw no coupe glasses or brightly coloured cocktails garnished with little fruit twirls. Dark wood and sticky linoleum floors, everything brown, including the liquid in the glass I was handed.

I held it away from my body, examining it.

Sensing my apprehension, Katy leaned over. Rolling her eyes at my snobbery, she said quietly, 'Just drink it.' I took a large gulp and my mouth filled with chemicals, the taste of hairspray. Ignoring laughter from those around me, I forced myself to swallow. Beth, in a display of skill and dedication, downed her brown liquid in one and told me I would get used to it. A little later, when I thought no one was looking, I whispered to Katy, 'What is this?' She took my glass and sipped the contents, handing it back to me with pink lip gloss where her mouth had been. It was, she finally decided, a double vodka and Coke.

Such drinks were generously donated by men who found great novelty in getting children drunk. Two glasses and thirty minutes later, my cheeks were warm and I had forgotten I was afraid of being discovered in underage deceit. Now I happily accepted any and all drinks, not bothering to ask what they were or who they were from. My eyes and legs were more unreliable than they had ever been before, and I leaned against a regular called Donald, talked to everyone a lot and easily.

My vision was blurry, but I remember clearly how I felt. For the first time in my life, I felt free from myself. Less of a shedding, more of an abandonment. And I loved it.

•

In the years of high school that followed, I didn't drink responsibly, nor was I encouraged to. Drinking was important, a competitive extracurricular activity. On each weekend trip into town, I thought about my next drink before I'd finished the one in my hand, and I pushed my body to reveal its weakness in streams of highlighter-coloured, alcopop vomit.

Wasted at Katy's house party, I'd been calling every taxi service in Thurso for at least an hour. Resigning myself, I clumsily squeezed past straggler couples still kissing in the hall, stumbled through an open door, was pleasantly surprised to find a coat-covered bed, and after clearing a space, immediately passed out on it.

Waking in the early hours I felt a warm, thick hand in my pants, touching my vulva. Brain flooding with information, I noticed that my trousers had been pulled down around my knees and what felt like an erect penis pressed into the back of my leg. Too scared to turn around, I lay still and prayed that it would stop. Time passed. My eyes flickered open once, and I saw that at some point Laura had joined me in the bed and was still asleep, our noses just centimetres apart. If I could get through this quietly, I told myself, no one would ever have to know.

But my body had other intentions. As the hands began to remove my underwear from my hips, I flew up into a sitting position and kicked the duvet from my legs. Scrambling across the mattress and onto the floor, I only turned to look at him, still in the bed, face turned away, once I had crossed the threshold into the hallway. It was Craig, Gillian's boyfriend.

Since I had been drinking, friendship had come more easily,

but I understood I was on the outskirts of stronger connections. I was sixteen and had to tread very carefully even to remain on the periphery. As I locked myself in the bathroom, I felt deep dread. I knew Gillian would see this as Craig liking me, me liking him. I knew she would wonder what I had done to elicit such attention. I knew the fallout would damage my tentative social standing. And, I heard Gill's voice in my head, *why didn't you stop him?* No matter how I answered, it was going to count against me. I made a pact with the bathroom mirror to keep my mouth closed.

Years later, I saw Craig opposite the Cameo cinema in Edinburgh. A deeply average man, headphones in, purposefully making his way down the street. My heart jumped to my throat as I watched his journey between the passing bodies and buses. I was shocked by how much the sight of him brought back. The shame. The guilt. The fear. I was right there again, unable to move. The body remembers.

•

I was five and had found it near Hamish's shop on the way to school. Sparkling blue and silver on the flat grey of the pavement; I picked it up immediately. It was a tiny piece of tinsel, and a broken plastic loop above the pretty metallic tassel made me think it had fallen from something much larger. I put the treasure in my coat pocket and forgot that it existed.

Returning home that evening, I dumped my coat near the door and ran upstairs to take off my uniform. At this time of year, it was a pale yellow gingham dress with a collar that chafed my neck, and although the mirror showed my skin unbroken it felt like it had been rubbed clean off. Hearing my mum's voice from below, I ignored the first few shouts of my name. Then, when her feet sounded on the stairs, I ran towards her like I had been meaning to the whole time.

Her face was fizzing. Lips curling over her teeth. Holding her hand towards me. 'What. Is. This?'

I saw my treasure from that morning, but her question confused me. I told her I didn't know what it was.

Her face fizzed a little more. 'You don't know? You stole it from the shop, didn't you?'

'No! It was on the ground! I didn't steal it!' I was shouting now. I had always been a stickler for the rules, and she knew that. I couldn't even play a board game without getting antsy about everyone doing it right.

But that counted for nothing. Instead, she grabbed my arm, pulling me down the stairs, my feet, in tights, slipping on the speckled carpet pile, gingham dress still scratching my neck.

'I'll be back. Stay here!' she shouted towards my brother who was building with Stickle Bricks in the living room.

She pointed at my shoes, waited impatiently as I put them on, then shoved her arms into her coat. With keys jangling in her hand, she pushed me out the front door and gestured down the steps.

'Where are we going?' I asked, terrified, grabbing on to the peeling black railing.

Shrieking, she worked to prise my fingers from the metal, 'To apologise to Hamish. I'm so ashamed. As if we don't have enough to worry about!'

On the street I kept my head lowered, stout tears clinging to my face as my mum marched me towards the space between the garages. A shortcut to the shop. I hadn't done anything wrong but I felt guilty anyway. And I knew I couldn't face mean old Hamish.

I yanked my arm from her grasp and sat down on the steps with my back against the wall, pushing my weight into the ground.

'Get up. Now.'

Shaking my head, sobs a hiccup in my throat, I crossed my

arms and resisted her attempts to lift me. She began to struggle. The keys, still in her right hand, dug into my forearm. And then she slapped me across my face. The world split in two. Before and after. I looked at her and saw that her eyes were filled with tears. The brown harling of the building pushed its points into my back.

Now the blows were coming at any part of my body that she could reach. I curled over, small as a garden snail, until she wound all ten fingers into my hair. The terror as she pulled me in every direction. The relief when my head did not hit the wall.

And then she let me go, and began to walk back towards the house, crying, repeating, 'What's wrong with you?' From my spot on the ground, where rough wall met smooth concrete, I didn't know if she was talking to me or herself.

•

As I got older, my mum would produce multi-page, handwritten letters on scruffy-edged paper, carelessly torn from a wire-bound notebook. The bigger her outburst, the more shouting and hitting, the fatter the envelope. She was sorry but what could she possibly do when I was so troubled and she had sacrificed so much already. I hated the way she wrote. Emotionally lavish language interspersed with quotes from songs she had been listening to when she wrote it. Probably drunk. The words she shared meant far more to her than they did to me. Each page had a 'PTO' at the bottom, rendered useless by the fact that once you had read one of these letters, you had read them all.

Even after it was over, I had no one else to comfort me. She wanted to, she was always sorry, but I couldn't find solace in being physically close to my mother. Her hugs left my body rigid, shrivelled, unable to forget what had just happened. If I could, I slipped away, and without exception, every time I remember feeling at ease or peaceful, I was by myself and I was outdoors.

Loch More was one of these places. The land, already sparse, became barer as I approached the dam. The water was surrounded by moorland, except in the far distance where lines of moving clouds were bent to the will of higher ground. Silence was disrupted only by the sound of small, wind-driven waves, and the shifting of fine-pebble sand under feet.

Beside the dam was a small house with derelict outbuildings, and behind those was a ledge of rough-cast concrete that sloped gently into the loch. To get to this ledge you had to squeeze down the side of the ruins or jump a wall, careful on landing not to gain too much momentum and end up in the tea-coloured water. With the stone wall behind my back, water lapping at my shoes, not a soul could see me.

Even on a day when the wind was howling, I would disappear past that old building, zip up my jacket over my chin and tuck my knees in as tight as I could, so the water, egged on by the gale, couldn't quite reach my toes. I'd stay there until my mum's attempts to call me back changed from singsong to staccato. I felt sad to leave that place and whatever it was that it made me feel.

Years later, undertaking my second attempt at my first year of university, I would receive a call from my mum saying she had organised, paid for, and would enforce, a hypnotherapy session to deal with the damage done by an abusive boyfriend (mine, not hers). Soon, I was lying on a foldaway bed in a poorly finished bungalow extension, eyes closed, scepticism coursing through my veins.

I was so reluctant to take it seriously that the therapist had to ask me multiple times to squeeze my hand into a fist and imagine the time in my life I had been happiest. I felt I could reasonably be expected to do one or the other, but not both. Finally doing as I was asked, I was embarrassed to feel tears prickling behind my eyelids as I was transported back to that ledge on the water at Loch More.

Stripping Out

*Offering their leaves to the rain, the trees above create a soothing soundscape. I'm grateful for their shelter, not because I mind the weather but because mud makes everything harder. Thus far, the ground beneath us has remained solid. Crouched in waterproofs and a wool hat, I pull the old wall apart, and as its stones fall on the grass around me, I feel my frustration building. A core rule of drystone is that every stone should go lengthwise into the wall. But each stone I remove doesn't follow this directive. Every single one is traced.

I ponder the reasons why someone would put in a whole course this way. I try to be generous. Maybe they were inexperienced. Maybe they didn't have what they needed. Magnanimity slipping, I wonder if they cared at all.

Once you've acquired the basics, the best way to solidify your understanding of drystone is by taking down broken walls. Counterintuitive: that dismantling a wall can give a solid understanding of how to build one. But in those early stages, it's hard to imagine the long-term effects of any of your decisions. Most seem inconsequential. It's near impossible to visualise the monumental obligations on one stone over decades, or how a waller's impatience can mean the difference between a wall that lasts three hundred years and one that lasts just thirty. So drystone repairs offer the gift of learning through the mistakes of others, a snapshot of the many lessons taught by time and the relentlessness of fundamental mechanical forces.

My piles of stone have begun to merge, so I take a break from stripping out to organise. Sorting your stone will often be enough to understand why a repair is necessary. Looking at my builders, I sigh. I'm going to need more. I'll bring them, load by load, in a wheelbarrow across this boggy field. If I built back using only what I had, the wall would eventually encounter the very same problems that I am here to mend.*

Among the crowds walking between the grand sandstone buildings of King's College in Aberdeen, I carefully enunciated every word. I swapped my glottal stops for hard Ts, hoping an accent that was hard to place would imply advantage and position. I leaned into my light skin, letting those around me see what they expected. It's not that I didn't want to be part Pakistani. It just felt like I was in a tight spot. Yes, the racism I'd experienced in high school had cut deep, but also, I began university on the 12th of September 2001.

Pairing socks, I'd watched the Twin Towers fall on the TV in the corner of our living room. The sound was turned down, and while I had some understanding of the gravity of the situation, I had a lot to do. My mind was at least a day ahead, mentally mapping Aberdeen, playing out scenarios. Imagining the possibilities of my new life.

In the hours after the attack, the scale of the tragedy became obvious. Sobered, I watched nervously as 'anti-Muslim' sentiment gathered pace. Although I'd grown up being called 'Paki', I'd learned that this word has little to do with actually having family from Pakistan. 'Paki' is a lazy, catch-all term. I knew enough to understand that 'anti-Muslim' was going to translate into anti-brown people, and anti-brown people would turn out to mean anything 'brown', like the pagri that Sikh men wear, or a name that isn't white.

As a child, I had delighted in spelling my mum's family name in my head, saying it fast, the way we did with 'Mississippi' in school. M-O-H-A-M-M-E-D. Sometimes I said it out loud when I was alone, enjoying how full the sounds felt in my mouth. I'd never been comfortable with my father's surname. It was a reminder of a man I did not know, who did not know me. I'd considered changing my name to 'Mohammed' many times, in recognition of my mum and her family. The people who knew and loved me.

But on the 11th of September 2001, I knew that becoming

Kristie Mohammed would open me up to a whole new world of violence. Fair-skinned privilege let me moonlight as a white person. And while I was mostly happy to do that, I was ashamed, too. There was a lot of shame.

Able to reinvent myself, separate from the past, I was telling so many lies and withholding so much truth that it became difficult to keep track. The comfort of being liked turned sour when people asked about high school or my family. The thrill of being told I was pretty was dampened by tilted heads and questions. 'Are you . . . Spanish or something?' There were good people. But there's no way to make real friends when you are lying about who you are. So I was lonely. I was vulnerable.

Daniel was militantly confident. That first week, walking to town with our flatmates, he pulled me behind a building, shoved me against a wall, put one hand around my neck, and kissed me. I tilted my face toward his. He wasn't conventionally good-looking but a lot of girls seemed interested in him. And he had chosen me. When he told me not to tell anyone else 'because they would be jealous', it made sense.

Even after we had been sleeping together for months, Daniel held me firmly at arm's length, frequently reminding me that I wasn't his girlfriend. Remembering the times he'd cried in my arms, I secretly disagreed.

In his room, waiting for him to return from the bathroom, I paced, picking up small objects from his shelves, turning them in my hands. A cheap five-a-side football medal, a torn-edged ticket from a foam party. Working my way towards the window, I saw a piece of paper on top of his stereo. I picked it up. A hand-drawn spreadsheet.

At the head of each row a woman's name. Lauren, Emma, Hazel, Kristie, Gemma, Annabel, Sophia.

Some of these were my friends; one was the expensively tanned girl who lived in the flat above Daniel. At the top of each column was a category. Face, arse, tits, body, hair, personality, humour, and, as he ran out of space, he had written 'SM', which I read as 'smile'. Each girl had been scored from one to ten in each class, with their total added up in a separate and final column. Reading through my scores, I saw that I'd lost points on 'body' and on 'smile' and had the lowest score for 'hair', just four out of ten. I made up for some of those lost points with humour. Fourth overall.

When Daniel walked back into the room, I still had the paper in my hand. With a smirk teasing the corners of his mouth, he took it from me, put it back on his desk and busied himself with his laundry basket.

I wanted to play it cool but the voice I heard was small and hurt. 'Body, I get, I suppose, but . . . Hair? And you said you loved my smile.'

A snigger skipped out of his mouth and, meeting my uncertain gaze, he seemed happy to inform me, 'SM is not smile, Kristie. It's *smell*, and the hair, well, that's about . . .' He gestured at my crotch with his eyes. My cheeks burned.

Perhaps sensing that he had gone too far, he pulled me onto the bed, tickling me until my face softened. Lying there, I laced his fingers between my own and focused on our closeness. The breeze of his breath on my skin. When I shut my eyes, the shame could exist in a different part of time.

I felt him move closer and, between small kisses, he asked, 'Do you think it's being part-Pakistani that makes you so . . . hairy?'

He continued to kiss me as I wondered if I ever had to open my eyes again. He dropped his hand from my chin and I felt the bed moving. I listened to the sounds of rummaging, drawers opening and closing. Then I felt his weight back on the mattress.

Finally giving in and opening my eyes, I saw him kneeling, his arms extended towards me, both hands held flat, drolly proffering an orange plastic razor.

He was excited, thrilled even. 'Get that score a bit higher?'

Daniel shaved my pubic hair with a Bic razor and an unsteady hand.

Afterwards, he remained between my legs, running his hands over his work. When that novelty had worn off, he bounded out of bed and leaned over his desk. Spinning back around, again pleased with himself, he presented me with the paper, and I saw that he had crossed out the number four, replacing it with an eight.

The thing about abusive men is that they're only pushing a more extreme version of what society tells women every day. At first, you sense absurdity. But when you look for reassurance, you see the same messages, more quietly spoken, everywhere. When you ask the people around you, they sort of shrug their shoulders, somehow unable to share your incredulity. So the abusive men become bold truth-tellers. Simply saying what no one else will.

I'd learned to ignore my own feelings, been taught that the people who love us hurt us, that I should always forgive and forget. I believed there was something wrong with me, and if he could overlook my defects, then I should do the same for him.

It's hard to distil how I felt during this period. But if I focus in and let the past wash over me, I can feel vibrations of desperation rising in my body. Desperation distorts and disorients. It leaves you without a working compass, looking for signs, asking for directions. Daniel gave directions like he owned the land and everything on it.

•

For Easter term break, I was heading north to see my mum. Daniel had agreed to come with me and I was excited. It felt special. Finally, a commitment on his part. The journey was seven hours with multiple bus changes, so we had to leave early. I arrived at his room at 6 a.m.

With both hands and a shoulder against the door, I kicked

my suitcase through the opening. Peering into the darkness, I saw Daniel in bed. Next to him, tousled blonde hair.

His nonchalance set the tone. He made no attempt to move as his pretty, wide-eyed companion sat up next to him, bed sheets tight against her breasts.

I looked at them with no idea what to say or do.

Breaking the silence, Daniel was smug. 'Since I'm a little, uh, busy, could you chuck some stuff in a bag for me?'

She stared at him open-mouthed. I stared at him open-mouthed. He continued: 'Bag's in the bottom of the wardrobe.'

Curtains closed, the room was filled with a grainy, artificial twilight, and I struggled to see what I was doing as I felt my way through a tangle of shoes and belts. Finally, I found his football kit bag squashed into a back corner.

'This?' I held it up.

He nodded slowly, like I was just beginning to understand something he had known all along.

Back turned, pulling socks out of his drawers, I could hear them whispering.

I turned around and looked at the girl, asking in a voice too high, too keen, 'So are you at the uni?'

Caught in the spotlight, she was startled into answering, 'Oh . . . no, I live in Banchory.'

I busied myself with filling an end section of the bag with boxer shorts. 'Oh wow, that's pretty far,' and before I could stop myself, 'How did you guys meet?'

It was Daniel's voice I heard, irritated. 'She's a friend of Dave's . . . You know what, can you just wait outside?'

From the other side of his door, I heard them laughing. I imagined they were laughing at me, the creep they had to ask to leave. Silence now. I wondered if they were kissing. Tears forming in the corners of my eyes, humiliation lumping in my throat, I told myself to hold it together, not to dare fucking cry. My face was tight and my breathing deep, as I held my eyes open, fighting the stinging sensation until it passed. When I heard the door open behind me, I wiped the evidence of my feelings from my face.

In the cold, fluorescent light of the hall, I saw his neck. Violent magenta bruises covered one side. In milliseconds my brain had run through all scenarios and made a decision. Rather than admitting the truth, I would face the full force of my mum's disgust, letting her believe I was responsible for this tasteless act of intimate branding. The tousled blonde's love bites would become my love bites.

Trying to pre-empt the next stage of this awful scenario, I turned to her and asked, 'Do you want to get the bus with us into town? We're heading to the bus station anyway.' I made sure to put emphasis on the 'us', the 'we're'.

Daniel's 'Yeah, that was the plan' made me feel like a groupie, a hanger-on, someone who wanted to be in the loop but wasn't.

On the bus, Daniel looked at his phone, manspreading in the space between his conquests. I sat quietly and tried to mentally hunker down for the long journey ahead, pushing out images of Daniel and Miss Blonde Banchory with fear of how my mum would react when she saw the marks on his neck. She'd never forgiven me for losing my virginity to my boyfriend at fifteen – something she'd discovered by opening a letter I'd written to him, and punished by grounding me in the bedroom for three months. I'd never spoken to her about anything to do with sex, relationships or intimacy again. Remembering this, I felt defiant.

When I'd left for university, I'd hoped the distance would be good for the dynamics between my mum and me. Give us a break from picking at old wounds. But I am a natural-born hedonist. And having grown up in the middle of nowhere, I was keen to experience all a city had to offer. My mum's parenting style was claustrophobic and now with more space to explore, it was clear I had a problem doing anything in moderation. It was everything she had feared.

In 2001, a drink was fifty pence at Aberdeen freshers' week. Shots of vodka in tiny plastic cups with just a bit more than a dash of mixer, face-twistingly disgusting. Upon walking into a venue I would order five from the bar and drink them back-to-back, like a smoker lighting a fresh cigarette from the end of another. Often too hungover to go to class, I'd sleep until it was time to go out again. I spent all my money on partying, clothes my mother would never have let me wear, and a fancy salon haircut, short, the way I had wanted it for years.

Now I was hundreds of pounds into overdraft and hadn't been to a class in months. I'd called my mum, but I hadn't been able to explain myself, and she had come down on me like a ton of bricks. When she'd refused to transfer more money, sending Tesco vouchers instead, I'd used them to buy bottles of Lambrini and Grant's Vodka. Part of me was too ashamed to admit to her how bad things were with Daniel. On the bus now, he was laughing with his blonde companion, and I wondered if the mess I was making could all be his fault.

When we arrived at the station, I hung back, leaving a respectful distance between myself and the couple as they said their goodbyes. Looking at the timetable boards, I narrowed my eyes and moved my mouth as if reading the details aloud. I hoped that anyone watching would see nothing but a young woman focused on the task at hand. I boarded the first bus north, and waited, swallowing to rid my mouth of the saliva that filled it again and again.

It felt like a victory when I saw him board the bus and make his way towards me. He stood stone-faced, waiting for me to move so he could have the spot by the window. When he sat down, he turned his head away, and we watched the city change to countryside in silence. I tolerated that for as long as I could, which was not long enough.

Voice shaking, I asked, 'Are you going to see her again?'

His head was still turned, taking in the views.

Swallowing again, I asked, 'Did you get her number?'

I craned my neck to see if he had heard, repeating, 'Did you get . . .'

Spinning round to face me, he hissed, 'If you don't fucking shut up, I'm going to get off this fucking bus at the next fucking stop.'

I flinched. I'd only been seeking reassurance, wanting him to promise he still loved me the most. Aware of the passengers around us, I tried to explain, quietly scrabbling to appease. 'Daniel, I . . . I . . . want to know, like, we're going to visit my mum, I need to know . . .'

When Daniel picked up his coat and stood over me, my brain became childlike. Scared and fumbling to understand. Then, when the bus began to slow and I heard the driver announcing the next stop, he pushed past me to leave.

The horror. I lost all sense of our public surroundings. Lawless, I grabbed at his hands and, when he shook me off, I clung to a fistful of his T-shirt and the back pocket of his jeans. I cried. I begged. How could I make this stop? Racking my brain for the magic words, I managed, 'Daniel, I promise. I will never talk about this again. I promise. It's done. It's over. I'm sorry. I promise.'

Still standing over me, he waited until the doors closed and the bus pulled away from the kerb before sitting back down.

The visit was a disaster. To turn up alongside an arrogant, callous boyfriend covered in love bites looked to my mum like another act of blatant disrespect. Daniel was angry that she insisted we sleep in different rooms, and, used to adoration on demand, he took great issue with her disdain, threatening to leave if she didn't stop not adoring him. Just a few days into the trip, standing on the empty platform at Thurso's small station, I watched his train leave. I had never hated my mum so much.

•

I want to enumerate all the things Daniel did to me over the next two years. To create a tangible record, outside his control, here on these pages. And I would, if words would not so spectacularly fail me. Each adjective feels puny, simple chronology inadequate, to capture the truth of an experience that destroyed me in every way a person could be. The world had already done a pretty good job of convincing me that I wasn't worth much, and Daniel, with unwavering commitment, finished that task.

My second attempt at university culminated with me three months pregnant. Daniel and I had spent a lot of time together since I'd shown him the test's second faint red line. So I was surprised at how sullen he was on the way to the hospital. Wasn't he pleased I had finally agreed?

We had our own room on the ward, where Daniel sat in a chair, again ignoring me to stare out of a window. Intense pain filled my body, and I curled up on the plastic-lined mattress, starched sheets cold and stiff against my skin. I spoke when cheery nurses came to ask questions, or to check the cardboard bowl in the toilet that was slowly filling with blood and lumps of soft tissue.

The hours passed slowly, and I was relieved when a middle-aged nurse with an English accent told us that we could leave. She explained that although the procedure was over, and they had no reason for concern, I should have someone with me overnight in case of complications. Daniel agreed; he was friendly, benevolent, telling her I was in good hands. He carried my bag to the taxi.

On the way back, I felt the future settle in the pit of my stomach. He still wouldn't talk, and now there was a finality to his silence. It was when the taxi pulled away from our halls of residence, leaving us standing in the rain, that Daniel began to speak.

He told me that he hated me. That his kindness over the last few weeks had been to ensure that I said yes to the abortion.

That he came to the hospital only to make sure that I went through with it, and that he would rather die than have a child who shared my flawed genetics. It was over between us, and he would never speak to me again.

I could do nothing until he turned the corner past the laundry rooms and disappeared out of sight. Then I fell to my knees, feeling blood seep from the too-small sanitary pad and onto my trousers. Hands on the cold concrete of the residence steps, I vomited.

•

Overcome with grief, I kept the times I left my room to a minimum. Seeing other people meant performing and I couldn't. I waited until my flatmates shut their doors and turned off their music to make my dinner. I stretched my tiny budget by boiling two potatoes and, while they cooked, frying half an onion in a pan. When they were done, I poured some hot water over chicken gravy granules, then mixed the lot together. The smell made me think of roast dinners at home.

Yorkshire puddings pulled from the oven still sizzling and popping in oil, my brother negotiating with my mum how much cabbage he had to eat. A game of Frustration afterwards in the living room, the resistance of the clear dome against my fingers and then the sharp click as the dice jumped inside the plastic. Before bed, Mum brushed the tats out of my waist-length hair as I complained and wailed at the injustice. The tenderness of her 'I love you' as she switched off my light. If I called her now, she would hear the truth in my voice.

When my dinner was finished, I washed the dishes and went straight back to my room.

•

From my window, I got to know people I'd never seen before. The tall guy who carried a unicycle over his shoulder, the girl who put seeds on her windowsill for the birds, and the other girl, dreadlocked, who practised backbends and hula hooping on the grass. Although I wouldn't admit it to myself, I was looking for Daniel. Hoping to catch a glimpse of him among the comings and goings, I diligently watched groups heading out to the pub and then returning later in smaller numbers. I saw those who liked to plan ahead using an empty launderette at 2 a.m., their sorting and folding a late night stage production illuminated by the lights overhead. I watched, and life continued without me.

•

I'd been bleeding on and off since the abortion, but this was different. Sitting on the toilet, I felt something large move in my abdomen and, squeezing my pelvic floor, I stopped its progress. Body tense with knowing, I willed it back inside me until I could hold it no longer. The object fell with a splash into the water below me.

The paramedics helped me up off the toilet, each one with a hand in an armpit. When I was fully standing, one of them leaned down and awkwardly wiggled my pants and trousers back to my waist. I lifted my arms away from them to show I could stand by myself and they exited the small space of the bathroom to wait. I buttoned my jeans slowly, taking a moment to prepare myself before turning towards the toilet. I told myself I'd look once and leave.

The foetus that lay in the water looked like a bloody mango pip. I had imagined everything, expected anything, but not that. The fluorescent light, the crimson kernel, the awful surrealness of it all. An almost-baby, Daniel, me, my flawed genetics. Keeping my promise but fighting tears, I reached over to flush before closing the door behind me.

At the hospital, a nurse told me what I already knew. Despite my difficult decision, blood and pain, the abortion had been incomplete. Something else that I, and my useless body, couldn't get right. Everyone else just had an abortion. And it was fine. Not me. I had to have a miscarriage, too.

Another nurse told me that I didn't have to stay overnight, but it was probably best to have someone with me in case the bleeding worsened. My heart sank. I'd hoped to stay at the hospital. I had no money, not even enough for phone credit, and even if I did, I couldn't think of anyone to call.

During my time with Daniel, I'd lost a lot of friends. He had convinced so many people that I was insane, and others were repelled by the fact that I always seemed to be in crisis. Of course, this situation would convince them further. I lied to the nurse and said I'd call a friend when I left.

Standing in the hospital foyer, I looked out at the busy street in front, reluctant to return to the world. A kind-looking woman with long dark hair walked by, a paper-wrapped bouquet in her hand. Shaking, I approached her and asked if I could borrow twenty pence to use the phone. She stopped and fished for her purse, looking concerned as she passed me the coin. Ducking her head slightly to make eye contact she asked, 'Are you OK?'

'I don't know,' I said, angry, and turned away before the tears started.

When my mum picked up, I lost the ability to speak. Instead, I gasped and gulped through sobs. I was still trying to form a sentence when my money ran out. I hung up, and as I prayed that she would call back, the phone rang.

'Mum?'

'Kristie, is that you?'

My voice cracked. 'Yeah, it's me. I'm . . . I'm not doing well, Mum.'

'What's happened? Tell me what's happened!' She sounded frantic.

I talked for a long time. The abuse, the abortion, the miscarriage. When I stopped, there was no response. I tried to slow my breathing so I could hear. 'Mum?'

'You were pregnant?' Now her voice seethed.

I heard parts of what she said after that. A woman who sounded an awful lot like my mother called me a whore. The blood rushed loud in my ears, and before I knew what I was doing, I'd slammed the receiver back into its cradle.

Later that night, I carefully dismantled a razor. I separated the blades out into a cup and poured hot water on them, washing off the soap and hairs by hand. Holding one tiny grey blade between my forefinger and thumb, I steeled myself for the first cut. Scared, but focused. Starting at my knees and working my way up to my thighs, I moved slowly and deliberately, cutting myself in thin, evenly spaced horizontal lines.

The blood showed first in turgid little drops, then, when their surface tension broke, it collected in small pools and trickled down my leg in a meandering, scarlet line. I grieved for my younger self. A razor concealed beneath bubbles, scared then, too. But now the blood was a testament.

When I was finished, I calmly called the residential assistant. Waiting on the edge of my bed, I let the blood from the cuts puddle and drip, and it felt like I was floating. Then people were in my room, taking the blade from between my fingers, collecting the others and putting them into tissues.

A guy, not much older than me, said that he had called my mum and she would be here in a few hours. Awkward, hovering, unsure of how to deal with such a distressed young woman, he asked if I still wanted to hurt myself. I reassured him with a quiet 'No'. When they left, I packed my things as best I could, flimsy dresses puddling on dirty trainers in a duffel bag. Every movement like a dream.

We drove in silence. I passed the time mesmerised by the ladder of incisions on my legs, running my fingertips gently over the beginnings of thread-like scabs. I spent almost two hours just loving them. Driving over the Kessock Bridge, I suddenly felt the need to share. Turning to my mum, I asked, 'Do you want to see?' Her face flinched, and I enjoyed that too. I looked right at her until she answered carefully. 'If you want to show me, then, yes, I do.'

Pulling up the material of my dress, I exposed the cuts. Darker in the middle and faded at their edges to a bright pink that looked almost neon against the pale skin of my upper thigh. They were the most beautifully indisputable things. She looked quickly before turning back to the road, staring straight ahead. I saw her eyes fill with tears, and I was angry, and I was glad.

•

In Caithness once again. My mum took her job as a prison guard seriously, limiting where I went, who I spoke to. Do the crime, do the time. I sensed that she was angry about everything and the only way to appease her would be to immediately get better. To her it was simple: I was too good for him and she hadn't liked him anyway. Plenty more fish in the sea!

I hated her and her wilful amnesia. Forget so she could forget. *Whore.* I couldn't. So I was trapped here, and alone here, and it was my own fault.

Looking for escape in the places I had found it before, I forced myself onto the land. Walking the back roads near our house, I faced the familiar long view across the fields. Amid so much space I wondered what it was that I could feel against my skin, constricting my breath.

A silent drive to Loch More. Mum walked the pebble beach as I made my way behind the outbuildings and onto the sloping ledge, tucked in my feet, stared across the water to the hills and felt numb. Any land I traversed, internal or external, was still the land of Daniel. So I trudged back to the car, rode home and let the pain bury me.

Under its weight, my body couldn't move. I lay in bed, chest heavy, arms, legs and mind deadened. Finally, my mum was sick of it. She began to shop me around to various clueless, often callous, medical professionals. Looking for a pill to cure this specific depressive episode, but if it could fix my inherent brokenness too, that would be nice.

Three doctors prescribed me antidepressants, but I resisted until my mum, desperate, gave me an ultimatum: take the magic pill or leave.

And then I didn't exist.

The pain had been unbearable, but it had been a pole star, and now thirty milligrams of Mirtazapine clouded that dark sky. For months with only sleep between me and vast nothingness, I spent my limited waking hours in a thick chemical fog punctuated by brutal brain zaps. Before the meds, suicidal ideation had brought tears to my eyes and pain to my throat, my body racked by the sadness of not wanting to live. On the meds I felt nothing. Life and death were equal prospects.

Despite my stupor, this terrified me, and one morning I found myself popping small orange ovals out of green and white blister packs, dropping them into the toilet before flushing. And as the pain returned, it brought direction.

Edinburgh 2004 was my third attempt in as many years at a first year of university and I don't think a single part of me really believed it was going to work.

Daniel contacted me within a week. He was from the city, and when he heard through friends that I was there, he called to say he would be in town. The next day. I went to meet him, my pubic hair freshly shaved, at a bar under the arches of the Cowgate.

Edinburgh is beautiful. Everyone knows that. A jostling, joyful congregation of five hundred years of architectural one-upmanship. Seven hills, seven views. The main peak of the city a watchful, endlessly patient volcano. Delicate scales of ancient fish were fossilised in the Caithness flagstone that lined the streets I walked to meet Daniel. I was blind to all of it.

We got terribly drunk and stayed and stayed, ignoring requests to leave, until the bartender resorted to rudeness. It was summer, the night was warm and humid, and we walked the two miles back to my halls of residence, talking, sometimes laughing as we held hands to stop ourselves from staggering into the road. Back at my room, everything felt carefree until he kissed me. After he fucked me, I cried.

Feeling his body next to me restless with the obligation of staying the night, I willed the tears to stop. The harder I tried, the more the pain elbowed its way to the front. Sighing loudly, Daniel stood up, mouth set in a firm line. He picked up his jeans from the chair beside the bed. My sobs grew louder.

He dressed, laced his shoes angrily, and as he stood, fingers on the door handle, he turned to look at me.

'You know, I only did this 'cause I thought you'd be less crazy now.'

He left without closing the door.

•

I took everything I was feeling and transformed it into something else. Pushing down the self-loathing until all that was left was a general sort of intensity, I threw myself into the distractions that Edinburgh could offer. People, parties, pleasure. I experienced my first true alcohol-induced blackouts, discovered class A drugs, and fell headfirst into weekends of indulgence followed by weeks of crushing depression.

Those around me seemed able to balance fun with functioning. My friend Maria was a residential assistant, doing a PhD, teaching her hot, younger boyfriend how to orgasm through prostate massage, throwing a full paella together on a Wednesday night, and was the sexiest woman I had ever met. But no matter how hard I tried I couldn't keep up with the day-to-day demands of deadlines and reading lists. If I missed a class, the fear of having to explain myself, or catch up, developed into panic. I could put on my shoes and pack my bag, but not head out into the world to make things right.

This time I didn't have an abusive relationship to blame for my under-achieving, so I reluctantly accepted that this must be who I was. Just kind of sloppy. Reinforced when, again unable to curb my spending, I used up my student loans by December. I took work where I could but struggled with all the normal parts of a job: time keeping, responsibility, teamwork. I moved from bars to restaurants to door work and back again, usually managing just a few shifts before it all fell apart.

As the year went on, my avoidance turned to something more phobic, and even close friendships became distant. Asking for support would have meant a confession, showing them who I really was. But I felt I should be much more impressive to be worthy of their efforts. I lied a lot, and about a lot.

It wasn't long before I found myself without any sort of job, even deeper in debt, and unable to pay rent. It took being on the verge of homelessness, notes from flatmates angry about late payments slipped under my door, for me to call my mum (again) and ask if I could come back to Thurso (again). I threw my things into bin bags and washing baskets, and three days later, without really saying goodbye to anyone, I was being driven north once more.

•

Back in Thurso, another failure in my wake, I couldn't understand what I was doing wrong. For others it seemed as simple as wanting it and making it happen. I wanted the same things, but something always got in my way. With no idea of what it was,

I could offer no explanation, even to myself. My mum pushed for answers, increasingly frustrated at what sounded like the excuses of a person who didn't even care enough to tell a convincing lie.

One early morning, watching from our doorstep as the mist lifted first from the fields and then from the treetops, I decided that I must be a free spirit. What else could explain my inability to settle and find happiness where others did? I opened my laptop and carefully researched what it was that a bohemian like me should be doing. I thought about it, I read about it, and just one answer made sense. I was going to travel the world.

Mum, tired and worried, called her sister in Guernsey and asked if she could set me up with a job. At least there, saving to go travelling, I'd have someone looking out for me. When my mum silently handed me the receiver, I put it to my ear ready to say whatever I needed.

My aunt told me that the hours on the wards were long, that the work was hard and not for the faint of heart. That I would be in a position where people would rely heavily on me.

'Right,' I said, 'I can work hard. It sounds great.'

'No offence to you, Kristie, but it's – this is my job. My reputation.'

'I'm not offended,' I said, deeply offended.

My private indignation increased when my mum gave me the same warnings several times over the next few days. But I would have agreed to walking around on a dog leash. Anything to get me out of Caithness.

A brain injury unit for the elderly might not seem like the most obvious place for a bohemian to start their journey of self-discovery. Long, draining shifts. The warmth and weight of elderly bodies. Learning how to clean carefully behind foreskins. But in Guernsey, amongst the suffering of a woman who would grab me as I walked past, pleading for something but never able to say what, and a man whose face was raw from his own saliva,

I was the opposite of a burden. I was useful. After years of fucking up, I needed that.

•

When I'd first arrived in Guernsey, the too-warm air had pushed me towards the crumbling swimming pool behind the nurses' accommodation. I'd walked its edges scooping out fallen, floating leaves with my fingertips. Then, lowering myself in, I'd spent three hours alone. The cool water held me, and in the soft, marbled light, I'd felt something start to gain substance again. Like a cup finally being refilled.

Six months later, I was on the way back to empty.

Counting wound dressings in the supply cupboard, I mixed up two boxes, lost my place and suddenly felt unsure that I had it in me to step back out onto the ward. Head down, trying to summon the courage, I stared at the hard-wearing, green linoleum beneath my disintegrating plastic clogs. There had to have been a kinder way to get that lady to lie still in her bed.

A nurse came in without knocking and leaned past my shoulders to take something from the shelves. I sighed and lifted my head. I liked the work, and working hard, but I knew I had to leave the job and probably Guernsey.

Then I met Luke.

Most people, interacting with a dementia patient, or someone with profound autism, will feel unsure and awkward, struggle to relate. But Luke, a student nurse on placement, was able to connect with anyone. There was not one piece of him that appeared uncomfortable. Our only interactions had been work-based civility, but I liked him. Talking to elderly clients, he took their leftfield comments in his stride, made them laugh and relax with jokes about the weather, his hair, anything. He was respectful, gentle, and, at just eighteen years old, Luke radiated kindness.

I made eyes and waited for him to approach me. But when we were paired for the morning's toileting rota, he was nothing more than polite. He lifted the patient. I replaced the soiled pad with a fresh one, breath shallow to avoid the force of the smell. I caught his soft brown eyes for just a moment. Smiled. And a round, still perfectly formed, vaguely green pea fell out of the heavy pad in my hands. It rolled across the floor, jaunty with a little bounce, then settled against Luke's worn, brown leather shoe.

It wasn't my shit pea, and it wasn't my shit pea-infected shoe, but I was mortified. I heard myself snort, subduing something between a laugh and a scream. I was cursed. People who had witnessed such an atrocity could never, ever have sex. But when I looked from the pea to Luke, he was also losing it. Tears streamed down our faces as we worked together to get our patient back on the ward.

After that, we spent every spare minute together. My memories of those few months are a little hazy, but the haze is one of summer light, sea spray and ice creams on the Jerbourg cliffs.

Then I was pregnant, and for no real reason I felt ready to become a mother. Perhaps it was the ultimate form of acceptance. Luke choosing to have a baby with me. Choosing to be inextricably linked, despite my flawed genetics. Of course, I see this only in hindsight.

At the time, it all made sense in that beautiful, linear way that falling in love and having a baby often does. So, with the best intentions in the world, a round belly, and my dress a green and black oriental-inspired two-piece with an elasticated waist, Luke and I were married aged nineteen and twenty-one in a small ceremony at the St Peter Port court house. Joanna, our Jo, was born a month later.

I lived the first year of her life in something akin to a fugue state. The only reference points I have now are photographs. Things look fine. I see my daughter sleeping peacefully in her cot, a well lived-in flat, dark lines under my eyes, under Luke's eyes, but smiles on our faces. How effectively they portray a false sense of tranquillity.

I'd never been good at staying on top of household chores. As a child, my room had been so messy that I would regularly come across fruits, moulding or desiccated, that I had forgotten months earlier. At university, I'd slept among piles of clean and dirty laundry that stayed on my bed for weeks. I'd always relied upon the seemingly random moments when, gripped with intense panic, I would clean the whole house in a few hours. Not a successful strategy for motherhood.

Things began to pile up. There were babygros heaped in corners, dirty bottles in the sink, paperwork for the health visitor under a pile of dishes on the coffee table. Our home began to feel oppressive, every strewn object a reminder of my failings. I was overwhelmed to the point of paralysis.

•

Two days after Jo's second birthday, cosied in my arms on the sofa, she began to convulse. I watched helplessly as she stopped breathing and a delicate blue tinge arrived on her lips, a contrast to the redness of her nose, which had been snotty for days. Foam began to pour from her tiny mouth.

I screamed at Luke, 'Call an ambulance!'

He stared at me, all medical training lost to panic.

I screamed louder. 'Noooow!'

Pacing the living room, holding my child still blue and not breathing in my arms, I barely heard Luke tell me that the ambulance would be ten minutes.

'Ten minutes? She's not breathing!'

We lived three minutes from the hospital, so I told Luke to get the car. Holding Jo's body on my lap, I began to pray out loud: 'I know I don't believe in you but if she's OK, if you make her OK, I'll believe in you. I promise. Please, God, please, please, please.'

Luke dropped us in front of A&E, and I ran into the building, him shouting after me. Something about parking the car, where he would be. I didn't care. I was frantically looking for signs that pointed to where I needed to go.

A receptionist glanced up, saw me, a lifeless child in my arms, and asked, completely disinterested, 'Can I help you?'

'My daughter isn't breathing. She's blue. We need to see a doctor.' I knew I was shouting.

Frowning, she said, 'If you could fill in these forms,' and held a clipboard towards me.

'Where's A&E?' I demanded.

'It's right there,' she said, gesturing to her left, 'but you need to fill in the—'

I was already through the doors heading towards a group of medical staff standing near a large whiteboard.

The next part happened quickly. Someone took Jo from me, and I watched as they put her listless body on a bed and cut off her clothes. Luke appeared and answered their questions as I asked my own.

'Is she OK?'

No one answered.

I asked again.

Firm hands on my shoulders moved me away, and a voice told me to give the doctors space to work.

Like high waterline tidepools, hospitals are still but alive. Every light, every noise, is urgent and means something, but I could decipher none of it. We stood in silence, too afraid to acknowledge the moment out loud. Minutes passed. My mind churned. I had given her Calpol! Had we told them about the Calpol?

A doctor broke away from the bed and approached us. 'It looks like Joanna had a seizure.' My foreboding escaped as a groan. The doctor continued, tidepool calm. 'We suspect it was a febrile seizure, triggered by a rapid spike in temperature. She'll have a long sleep now, but will likely be fine when she wakes. To be on the safe side, we should keep her overnight for observation.'

Relief. My knees gave way and I began to sob. Luke patted my back, crouching beside me on the cold hospital floor.

Jo woke up six hours later, immediately able to totter around the ward and show the nurses her ballet. Her hair, still tamed in the smallest of bunches at the side of her head, bounced as she loosely pas-de-bourrée-d across the vinyl floor in her nappy. 'Look, Mumma!' And seeing all the life in her, I could think of nothing but her small body limp and blue in my arms.

When Luke offered to get my toothbrush and some spare clothes, I told him no. He should stay. Later, lying in bed at home, not there for my child, I was both checked out and totally consumed by panic. I fell asleep praying to a God I still didn't believe in, asking him to please, please keep Jo safe.

By the next morning an intense, unknowable fear had settled. I lay in the dark, knowing I should call Luke. But I couldn't. Even when Jo came home, I didn't leave the bedroom. I ignored my phone and snapped at Luke when he asked if Jo needed a coat for the park because of course she fucking did.

Through the coming weeks, the deep distress stayed. Soon it was New Year's Eve, 2007. Lying on the sofa with olives, Shloer and a modest cheese board on the coffee table in front of us, we watched my favourite Eddie Izzard stand-up DVD. Halfway through, when the Three Wise Men are getting baby Jesus gifts at a late night petrol station, I was overcome by a feeling of profound dread. I was going to die. I wasn't a bit worried, nor was I considering the possibility of death. Instead, I was absolutely sure that it was happening. I began to shake.

Panic attacks became my most regular visitors. Sometimes they were so bad that my hands cramped into rigid, claw-like shapes. Luke knew what was happening, but I hid my crooked fingers under cushions and blankets, embarrassed. Outside the attacks, in the better moments, my brain began to force upon me the worst things it could possibly imagine. Scenarios played in my mind, so graphic that my heart would race and I'd raise my hand, slapping my temples, hairline, forehead to force them out.

What if I smothered Jo with a cushion, pressing down until she stopped breathing, her lips blue again, skin waxy and pale under the fabric? What if I got so tired that I put her in the oven and forgot and turned it on . . . how fast does a baby die in an oven? What if I stabbed her, too much blood spurting from her tiny neck? Always death, always vivid. I hated myself and what I was capable of.

The more I hid them, the worse the thoughts became, and I began to wonder if suicide was the only way to stop me from doing something terrible. My brain obliged by setting out, in gratuitous detail, every possible way to kill myself.

One day, Luke gave it a name. Intrusive thoughts. He had learned about it while studying to be a mental health nurse, and I listened as he casually shared the very same ideas that had been torturing me. I sat still, staring at the smear of ketchup beside the scratched wooden leg of the sofa as he explained that intrusive thoughts were common in parents ('more than half of new mothers,' he said). I rubbed hard at the ketchup with my foot, enraged that he, and the rest of the world, had kept this knowledge to themselves.

I loved Jo, but my desire to protect and care for her didn't directly translate into the skills to do so. During her first years, I asked and asked but was offered no solutions. What is appropriate punishment? Does punishment even work? How do I teach my daughter to be kind without teaching her to abandon her own needs? Is screen time as bad as they say it is? Instead, I listened, as if on a loop, to some generic version of 'sleep when your baby sleeps'. Even my own mother seemed unable to elaborate on her assertion that 'things will be fine'. I was at sea.

Despite being good at doing what I asked, when I asked, Luke approached parenting without initiative. Sure, he gave Jo late-night bottles and changed her fullest nappies, but I longed for someone to tell me what it was that I needed to do, and when I needed to do it. Or at least someone who could do things without waiting for my command. Instead, Luke stubbornly relied on me, and I in turn relied on the reading and research I had begun to do every day. It was so important and I knew I was getting it wrong. Trying to hold the weight of raising a healthy, happy human when I myself was treading water.

Back then, Luke had a comically sincere way of accepting criticism. Eyes soft and wide, head bent occasionally in penance, he listened in absolute silence, never interjecting or defending himself. He took it on the chin. I'd never experienced anything like it, and at first I saw it as a man ready to make real change. A couple of years in, I saw it for what it was: an act of contrition. Luke simply suffered through the punishment, my anger. Once over, absolved, he never thought about it again.

I knew I should communicate my feelings clearly and calmly, but I saved everything up, letting it out when I was too furious to keep it in. Once I'd felt that release, I'd go back to collecting wrongdoings, Luke would forget about it, and we would repeat that pattern ad infinitum, resentment building on both sides. At the time I thought we were struggling to overcome complex marital issues, but the truth is we were too young to navigate a relationship alongside our personal struggles. Which were many.

After we split, Luke saw Jo sporadically. He retreated as far as he could.

•

Now a single parent, I lived each day at the edge of my nervous system. The psychological equivalent of hand to mouth. Fulfilling the needs of the moment with no idea what might come next, or if I could manage it. I spiralled. Our flat had been messy but now it was filthy. Dishes on every kitchen surface for weeks, piles of wet washing beginning to rot. I couldn't make a phone call or even fill in a form. And I spent money like I had some, finding my way into a £2,000 overdraft, bills left unpaid.

In moments of difficulty, we feel a pull towards our parents. Even if there's mostly evidence to the contrary, we still cling, in the deepest parts of our bodies and minds, to the idea that they might be the ones to save us. When people asked if I was close to my mum, I answered 'yes' with confidence. Sometimes we spoke three or four times a day. So, a few years after she relocated to Edinburgh, I moved with Jo to a small flat two doors down. I was desperate for help, and for the comfort that only a mother could give.

When it comes to practical things, my mum operates at the highest level of responsibility. She proved everyone wrong in their bigoted opinions of mixed-race people and single mothers – and of her specifically as a mixed-race single mother. Jobs held firmly down, home kept immaculately clean, money budgeted to within an inch. She couldn't believe the state of my flat, was shocked by the mouldy cups left on the living room bookcase and concerned about the nutritional value of Jo's frozen pizza dinners.

It took just one month before I had to physically remove my mum from my flat when she lost her shit, throwing insults and a glass bottle. But the worst of her behaviour never emerged with her grandchild, and my mum took her role as a grandmother seriously. She was brilliant. Carefree, fun, affectionate and held no aversion to glitter. Jo often asked to spend time with her nana, longingly pointing up at my mum's window when we walked past her flat.

We didn't always see eye to eye, especially on the more progressive aspects of my parenting. I didn't discipline Jo with smacks or time-outs, and sometimes she slept beside me. I was trying to be intentional.

One afternoon, I answered my phone in the wine aisle as I was trying to figure out which bottle of Blossom Hill was cheapest.

'Oh, hi, Kristie.' My mum sounded apprehensive. 'You OK?'

Frowning while balancing two bottles of too-pink wine in my basket, I asked, 'Are *you* OK?'

'Yes, we're OK. Everything's fine. I thought I should mention what happened . . .'

I felt it in my stomach.

'. . . Well,' she went on, 'I was reading *The Very Hungry Caterpillar* to Jo and we got to the end, you know when, after he's eaten everything, and the green leaf, he comes out and he's a butterfly–'

I interrupted her. 'I know what happens at the end of the book, Mum.'

She paused to convey her annoyance at the interruption, then continued. 'Well, I asked Jo, "Do you know what a chrysalis is?"'

I could hear her swallowing, and tried to do the same, but my mouth suddenly lacked moisture.

The weight of wine, tangerines and laundry detergent hung on my forearm, basket tilting. My voice got louder. 'Mum, can you just tell me? You're fucking scaring me.'

'OK. It's not a big deal . . .' she said about this very obvious very big deal. 'I asked Jo if she knew what a chrysalis was. And do you know what she told me, Kristie? She told me, "It's a little lady willy!"'

A pause.

Jo had a book that labelled female genitalia. I'd read that it would give her a sense of ownership and instil the knowledge that her body was normal, healthy, whole and beautiful. Good adjectives, shoot for the moon, etc., etc. She had studied it carefully, with the intention of a scholar. *A little lady willy.* I struggled to contain my laughter.

But Allison did not find this funny. She was telling me that she disapproved, didn't appreciate having to deal with this. She asked me why I was teaching my four-year-old 'stuff like that'.

I could have apologised but I wasn't sorry. 'Mum, when I was little, you called "stuff like that" my "front bum". It took me until I was nineteen to learn all the right names! Vulva, labia, vagina, CLIT–'

The line went dead.

•

Proximity to my mum offered me the most logistical stability I'd had in years. But it also pulled me back into the dark dysfunction of my childhood. I was a young twenty-four with a toddler, an ex-husband and a forty-five-year-old mother who regularly gave in to her demons. I painted hope over truth, immersed myself in what felt familiar and walked directly into the worst years of my life.

With the past snapping at my heels in more ways than I really understood, with no knowledge or tools to help me move forward, I found little point in trying to run. Better to lie down under the soothing weight of inevitability.

Not soothed enough, I drank a lot.

Vomiting on the floor of strangers' flats and passing out in their beds. Starting the morning puzzled while pulling wet clothes from the sink before remembering it was my urine in which they were soaked. Drinking alone. Tipsy wasn't sufficient. I always had one eye on the bottle, dreading that final, half-glass pour. Over time, one bottle of wine turned into two. I wanted to feel nothing but the room spinning and gravity pinning my body to the bed.

I lived for those moments of insensibility, preferring the powerlessness I'd chosen over that which felt forced upon me. Giving myself back to Jo and her one thousand questions in the mornings was jarring. Even more so through the low-lying, thick clag of my hangovers. My brain and I were on no better terms than we had been when Jo was a newborn. There was not a minute of my existence that I didn't hate myself.

It's a strange thing to hate yourself while trying to teach your child to love herself. But I did try. Jo was sweet, silly and sensitive. From an early age she committed to a deep dive on anything that caught her attention. X-Men, Pokémon, dinosaurs: she was a cold-hard-fact machine. If she felt she had hurt someone, she would ruminate over it for weeks, unable to cut herself slack. When she learned to write she would leave notes on her blackboard for me to find. *I love Mum she is nis Be Cus She is vere nis Be Cus she is nis.*

Late one night, I found her fast asleep with a drawing of a strange asterisk-shaped plant under her head. The next morning, she explained that she'd heard spiders were afraid of bamboo and, having never seen that plant before, she had drawn her own impression of it, in the hope it would keep the spiders away. Another night, standing at the living-room door, hands on her hips, eyebrows raised, doing her best impression of an adult, she announced, 'Well, this is a fucking mess.' I loved her, but I also liked her.

I had no idea how the world I lived in was affecting my behaviour, or how my behaviour was affecting Jo. And as my energy waned, intention turned to intuition: only ever a direct translation of how I myself was parented, or a reaction to how I felt at any given moment. Looking back, I see that I was in desperate need of care, understanding, guidance and rest. I had so little to give to my child.

As I had been, Jo was forced to navigate the highs and lows of her dysfunctional mother's moods. She subsisted on the rare moments of calm and focus that I could offer. A second to breathe before she was plunged into the depths of an outburst, followed by my inevitable retreat to self-pity.

Although I wish someone had told me that dating during this time was a bad idea, I also know I wouldn't have listened.

Every interaction I had with a man was like being back on that bus, Daniel standing over me threatening to leave if I didn't shut up. Men took advantage of my despair, and in each encounter I chipped away a little more of myself and handed it to someone who, as a rule, did not care to have it. Sex was something that happened to me. I passed every touch through a complex set of filters to determine appropriate and expected reactions. Reactions that I believed men would like. I dissected my physical experience until it existed as theory in my mind. Soon, no part belonged to me.

•

In the disconcertingly quiet after-hours of the bar, I perched on a stool and felt inebriation vertigo pull me towards the floor. Steadying myself by hooking my ankles behind the stool's thin metal legs, I watched him make the drink that I had already declined several times. His dark hair was gelled into a belligerent shape and the chain hanging from his jeans rattled occasionally against a fridge or the bar sink, echoing in the empty space around us. After a couple of minutes, he turned, presenting me with a tall glass of brown liquid. Leaning back against the till, he took a sip from his own short glass.

'What's this?' I asked, stirring the dark substance with my straw.

'A Conrad special,' he said, smiling.

I picked up the glass and sniffed at the liquid below the ice cubes. My nostrils constricted and alerted the rest of my body to the prospect of more alcohol. My stomach immediately mounted a raging nausea defence.

He moved away from the bar and sat on the stool next to me. As he pulled me towards him, I lost my balance and fell, ripping the knee of my tights on the rough wooden floor.

Swearing, standing up, I wobbled, and he grabbed my waist, laughing and saying, 'Whoa, you're really wasted, aren't you?'

He guided me towards a bench, settled me with my back against the wall, then lifted my drink, put it on the table and pushed it towards me.

I picked it up, braced myself, chugged it. After I swallowed the final mouthful, the shiver that moved through my muscles forced them to surrender. Hands too slow to save me, I tipped to the side, falling until the weight of my head rested against his body.

Pushing me from half past three to midnight, he put his arm around my back, hand on my hip. I laid my head on his chest, and wound my fingers into the fabric of his shirt to stop myself from sliding into his lap. Closing my eyes, I felt the hand of the universe spinning me into oblivion. From there, every memory is a snapshot from within that whirling dance. Flashes of faces, objects and sensations, everything else a blur.

Him hooking his arm under my waist, roughly pulling my slack, intoxicated body back to kneeling. Cushion tassels and coarse carpet. The bristling fibres pressing into my hands and knees, then, when he grew tired of supporting my weight, my face. The smell of dust, cigarette smoke and spilled drinks. Pain. Fear. The feeling of my own body pinning me to the ground even as I willed myself to move.

On waking the next morning, I felt half alive. Stiff and hunched over, I undressed slowly in the bathroom, the full-length mirror on the back of the door revealing bruises on my forearms, torso, thighs. Holding my clothes in my hands, I sat on the toilet and saw the carpet burn on my knees. The stain on my dress. It looked like the marks of saltwater after a trip to the beach. He hadn't used a condom.

I picked up my phone. Sixteen missed calls. I'd slept through numerous attempts to return Jo to my care, and from the increasingly angry tone of the messages, I understood that Luke had missed work. I texted to let him know I was up, and he appeared soon after, my mum in tow. Among empty bottles from the night before, both of them berating me, there was nothing for me to say except sorry.

Later, Jo asleep in her bed, I picked up my purse from the hallway floor. A small collection of pastel papers had been shoved into it. Smoothing out wrinkles, I smiled to myself as I recognised them from my childhood. I had two lilac fifties, a sky-blue ten, a pink five hundred, and no idea where they had come from.

Then, pieces of a memory. Working methodically to exit the taxi, falling, unable to stop myself. In one hand, my phone, and in the other a fistful of pretty paper notes. The taxi driver, seeing me drunk, dishevelled and sobbing, hadn't taken me to the hospital. He had handed me my change in Monopoly money.

•

After that night, I left Jo. I was unable to be her mother. Not only because someone else's needs were altogether too much, but because she deserved so much better. I moved into a flatshare and let Jo live with Luke. No, he wasn't capable of being a full-time parent but, evidently, neither was I.

Being a mum had been an anchor. Without that, I partied all the time, once again taking and losing jobs almost weekly. For months, I barely saw Jo. I loved my daughter more than life itself, but with a brain determined to return to the fold of chaos, chasing asshole men made more sense.

I preferred connections that were fraught and the tension of pursuing people I had to persuade. How I felt about these men was of no importance. The goal was never to like them, it was to make them like me. I immediately wrote off anyone who showed interest as obviously inferior, kind words repelling me the way cruel words should have.

So when I discovered that the guy I had been sleeping with was doing so behind his real girlfriend's back, I decided I should follow him on his 'round the world' trip. If I couldn't make him love me here, I would surely have more success in Australia. I bought a plane ticket and a loose white dress with blue trim. So boho.

•

As far back as I could remember, my mum had been searching for an answer to the question of what was wrong with me. The diagnosis that could explain it all. Sometimes she would float this idea gently, and sometimes she would shout it, telling me how hard it had been for her and my brother, how I'd ruined their lives. Hypoglycaemia, candida, bipolar disorder, schizophrenia and learning disabilities had all been floated as potential reasons for my dysfunction. So had selfishness, narcissism and being 'a bitch from fucking hell'.

We had visited specialist after specialist, at great cost to my mum, but the miracle diagnosis remained elusive. She had become more resolute in her quest and I had become more resistant. I was not willing to play any part in her absolution. I blamed my mum and she blamed me right back.

The only time she suggested therapy was when she was angry. Agreeing to a session felt like an admission, and I had no intention of pleading guilty. I was innocent. I wasn't going to voluntarily agree to them searching the house. They could come back with a fucking warrant. So when my mum suggested family therapy, I was amenable. A three-way split of the blame felt like a good deal.

Double-checking the address in the email on my phone, I walked down the concrete stairs to the basement level. Here, a peeling blue door sat ajar. Pushing it open, I was met by a wall of heat and the smell of damp. Inside, I followed echoing voices towards the end of a corridor. The room was stark. White walls, an electric-blue carpet (the hard-wearing kind you find in offices, sandpaper made of nylon) and red plastic chairs arranged in a circle. Already occupying the seats were two older men with hardback notepads and patterned woollen tank tops over shirts. Their blank stares prompted me to speak.

'I'm here for the family therapy session?'

'Ah, yes, yes, come in, sit down.' They spoke almost in unison, words overlapping. Not quite twins but absolutely twinning, the main difference being a pair of gold, wire-rimmed glasses on the shorter man's face.

I took off my duffel coat and smoothed my straightened hair. I could see out of the corner of my eye that it was beginning to frizz. Waiting for my mum and brother, we passed the time making small talk about the weather, and I reassured them more than once that the too-hot room was an appropriate temperature.

'Hello?' my mum's voice sounded from the hall.

Bustling into the space as she always did, my mother's aura was one of a hundred bags even though she carried just one. Dropping her tote on the floor, she noisily shook her way out of a green trench coat, unwinding her scarf at the same time. As she sat in the nearest chair, the one opposite me, she fanned her face with a hand and blew air from her mouth onto her top lip, which was glistening.

Already out of his seat, the taller of the two men moved towards the heater and, looking incredibly disappointed in himself, turned it down a few clicks.

Back with a clipboard in his hands, he drummed his fingers on its wooden back and asked, 'Waiting for one more?'

Reaching down into her bag, my mum disappeared behind her long, dark hair. Emerging, she fiddled with a tissue in her hands and responded cheerily, 'Just us today!'

My anxiety jumped from a five to an eight. This had all the makings of a trap.

The counsellors introduced themselves. Gold glasses was Stuart, and the other was Graeme.

Stuart, sitting up straighter in his seat, clasped his hands in front of him and began to speak. 'I've already spoken to you, Allison . . .' He lowered his head in a cordial nod towards my mum before continuing, 'but I think if you could give us all a sense of why we are here today, that would be great.'

Wiping her nose with the tissue, she breathed deeply in a particularly pointed way and then said, 'I've wanted to do this for a long time. As I explained on the phone, we need to get to the bottom of what is happening in this family.'

She looked over at the men, both writing in their notepads, and then continued: 'I'm treated so disrespectfully by my children and I'm the mug who keeps trying to fix things. It's always been like this . . .' She paused and wiped her nose again. 'Even with my own family in Glasgow, I've always been the black sheep, the scapegoat, used as a punch bag, sometimes literally, by a brother who called me a hoor for having children when I wasn't married.'

To roll one's eyes at the story of a young mother being punched in the face by her brother is cruel, but I couldn't stop myself. I'd known her brother had hit her. I'd heard the story so many times and even remembered her wearing a neck brace afterward, in so much pain she'd been unable to turn her head for weeks. What I didn't know was what he had called her. *Whore.*

Suddenly, I heard my mum's voice higher, more urgent: 'Do you see what I mean?' Emerging from my eye roll I saw that she was gesturing emphatically towards me while looking at Graeme and Stuart who were in turn looking at me, pens poised above paper. I was unable to take my eyes from the bracelets jangling on my mum's wrist.

Up on my feet, I knew I was shouting. I could feel the anger reverberating in my throat, but my hearing was dulled by a head so full of memories that my voice sounded quiet and far away. I don't remember what I said. With trembling hands, I picked up my coat and my bag and turned to face shocked male faces and my mum's tears. Then I walked out.

From the hall, I heard my mum's indignant yell, 'Do you see what I mean?', and as I climbed the dank steps from the basement, I imagined the two men writing furiously in their notepads, twin heads nodding. Yes, they saw what she meant.

•

I'd perfected the art of selective honesty from a young age. I was a bad liar but I could make others feel like I was saying a lot without really saying much. Over-sharing candid information about one thing, or one part of my life, helped people assume that I was honest about everything.

Alone in the difficult moments, holding all my secrets, I was often angry with friends, wishing they had tried a little harder, paid a little more attention, asked a few more questions.

Maria and I had been close at Edinburgh University. One of the exceptional people who had seen past the turmoil. We had stayed in touch, and I'd kept her updated with the same mass emails I sent to everyone else, including the one in which I'd casually mentioned my 'round-the-world' ticket. Basically, a holiday. I deserved it.

Even the first few lines of her email were difficult to read. Pulling no punches, Maria told me that she knew I lied a lot, and about a lot. My stomach clenched with shame, then anger. Outrage that kept me reading. Maria went on to say that she suspected this trip was about chasing a man who did not care about me, and that I was going to abandon my daughter, who so needed me, to do it. She told me I was avoiding things, distracting myself, squandering my potential. She was worried and I should let her know what she could do to help. I still don't know what persuaded me to peer beneath my initial fury to recognise something else.

Pivotal life events don't always result in hard and fast change. Instead, these crucial moments are an underpinning. Although I saw sense and didn't go on a 'round-the-world' trip, although I resumed an active role as Jo's mother, the rest of my life stayed roughly, chaotically the same. I still didn't like myself, pursued unavailable, arrogant abusers, drank too much, had an awful relationship with my mum and found the world entirely overwhelming. But by sending that email, Maria showed me that she could know the truth of my struggles and still see my worth. It's only in hindsight that I can say with certainty that her intervention quietly, lovingly altered the course of my life.

•

I wish I could say that I sought my own therapist for a healthy or noble reason. In truth, I wanted someone to be on my side. To nod their head. Say I was right, and my mother was wrong.

Having a therapist side with me didn't feel the way I thought it would. I was shocked when she used the word 'abuse'. First, I explained that there had been a misunderstanding. Then the guilt set in, trying to recall all I'd said, worrying that I might have been the one to use the word first. Defensiveness came next, anger that someone would say such a thing about my mum. Then shame at the thought of being someone who had been abused. Again, it felt like the world had split in two. Before and after. I left the session not knowing if I would ever go back.

After all, I'd got what I'd come for.

Once I'd acquired a therapist-backed assessment that everything was my mum's fault, I surprised myself by not immediately throwing that bomb in her face. The information felt dangerous, incendiary, and I examined it from a safe distance. I circled it again and again in my mind, trying to understand. What could it be for, if not to lob at the woman who'd raised me?

I did go back to therapy but, aware of the power it held, I skirted around details of my childhood. If we got too close, I would stop going for weeks or months. Then I could tell myself that therapy was kind of bullshit, anyway. I had expected it to help. Immediately. I'd looked forward to existing as healed, healthy and whole. But it was creating powerful tensions between the internal and external.

Like Maria's tough love, therapy breakthroughs resisted translation to my actual life. Sometimes I had no idea what to do with them. I was full of ghosts. Held in limbo, knowledge and insights needed to be put into action in order to be released from purgatory. Usually, I lacked the courage to do it. And, always, they refused to give me peace.

It looked like it all needed to be torn down. I knew I had a responsibility to make sure my trauma, my family's trauma, travelled no further along the generations. I didn't want to offer my daughter the same unsound legacy that had been given to me. A lifetime of repairing others' shoddy work. I wanted her to be able to rely on what I built for her so that one day she could build something entirely of her own. I didn't know where to begin. But I wasn't going to let Jo be chewed up by the world like I had been.

Sorting the Stone

As I climb the pile, stones begin to make their purpose known. Long, regular, flat shapes will work well in the cheekend. Stones that span the entire width of the structure might be throughs, and builders have at least one face that enables them to go lengthwise into the wall.

These stones were recently pulled from a field and a layer of mud conceals their details. I notice a good-looking oblong at my feet and pick it up, my gloved hands assessing its suitability. Not too big, not too small, no weird lumps or bumps and a face with a nearly ideal batter. Perfect. But turning it over, I notice a groove. A fracture. Running its entire length, this crack means the stone is likely to split once placed in the wall.

I'm not angry, just disappointed. I raise it above my head, and it drops with a dull knell onto the pile at my feet. It breaks along the fissure and in two other places where it was likely already weakened. I pick up three of the pieces and throw them into my pile of hearting, then walk to my snecks with the fourth and set it down carefully to keep it intact. Every stone finds its place. Nothing goes to waste.

Each, no matter how humble, is an integral part of the wall, and fulfils its own purpose while working alongside friction and gravity to form a structure that will stand the tests of time. The fate of each and the fate of the whole are inextricably linked.

I knew almost nothing about my father. Technically, our mum had left him when I was a toddler and Matthew was still a tiny baby. But he'd come and gone before that. As we'd grown up, she had spoken about him rarely, careful not to oust the fathers of our imaginations.

I'd taken details from the few photos I'd seen and run with it. Tall like me, jet-black curly hair and green eyes. He was a doctor, a lawyer. Crisp navy suit and a black leather briefcase with a gold combination lock, a code he told only to me. Married to a wonderful woman, soft in all the ways my mother was not. No kids, obviously. They wanted some. Wanted me.

At seventeen, being taken apart by Daniel, I'd seen enough movies to know that my dad wouldn't stand for it. He'd tell him to stop. Warn him, fist wound into the fabric of his fake Radiohead T-shirt, what would happen if he didn't.

Or he would just be nice to me. Take me out for lunch and listen.

It took me all of ten minutes to find his address through the electoral register. As I folded the single sheet of my letter into perfect thirds I told myself that it could be the wrong person, the address might be wrong, my note could get lost in the mail. If he didn't respond, it just meant he hadn't received it.

•

I saw them watching me from across the street. Sitting at a weather-worn table, alone on the pub's small concrete patio, they tracked my steps. I made a beeline, crossed before the traffic lights and waved. They were both in denim jeans with denim jackets, balding with shaved heads. I couldn't tell who was who.

One of the men stood up, smoothing his jacket, and I caught the first glimpse of my father. Shorter than I had expected, and nothing like the man I'd pictured. Not a lock of thick black curly hair, and little colour left in once bright green eyes. I also wasn't getting strong doctor/lawyer vibes.

The weight of the moment, amid fully realised social awkwardness, transformed his attempt at a hug into the reality of a handshake.

'Oh . . . oops,' I offered, taking back my hand and moving between stray chairs to sit at the other side of the table. I looked down, arranging my skirt underneath me, giving someone else a chance to start the conversation.

Shane spoke first. 'Ehm, this is your uncle Barry. Your mum knows him . . .' He shook his head. 'A long time ago.'

Smiling my brightest smile, I said 'Hi, Barry, nice to meet you,' then pushed both hands under my legs, letting their weight hold my fidgeting fingers captive. I hadn't known he would be here.

'Right, then. I'll get the drinks in. What are you having, Kristie?' Barry was on his feet, a leather wallet in his hands.

Noticing the high tide line of foam on six empty pint glasses in front of them, I said, 'A half pint of what you're having?'

Barry nodded, put a reassuring hand on his brother's shoulder, then turned toward the pub doors and walked away.

Shane spoke. 'I knew it was you as soon as I saw you.'

I looked up, surprised. 'Yeah?'

'Your walk is the same as your mum's.'

He pulled a small bundle of photographs from his jacket pocket. 'Your mum hunched over as well. To hide her height.' He shrugged his shoulders forward to demonstrate.

Spurs of indignation pressed my thoughts. This leap across the last sixteen years was unwelcome. Too familiar. I was angry that he would assume anything about my reasons for 'hunching over', especially since I didn't even know that I did it.

Gesturing towards the photos now in my hand, he said, 'I kept those,' and then his voice became too sincere. 'I always hoped I'd find you.'

I felt the prickling again, and swallowed to save us both from the words that were already in my mouth. *But I found you.*

I focused on the images.

The first photo, my mum, slim with high cheekbones and perfectly coiffed hair. Next, a baby who must be me in the arms of a severe-looking white-haired woman. As I glanced up, he was already answering, 'Your granny. My mum.' Pausing for a second to take in her features, I compared them to my own. Maybe something around the eyes.

Then, silence.

I held the photos as my mind raced to fill the pause. I told him, tripping over my words, how I'd always loved photography. How I'd spent hours in my school darkroom processing negatives. I told him that probably soon everything would be digital, a real shame. 'Like records!' he said, interrupting and loud. 'I loved records and now it's all C-fucking-Ds.'

The clink of glasses alerted us to Barry, back at the table, hands full. Extracting my half, I thought it looked mightily insufficient and wished I'd asked for a pint.

Shane and I chatted while Barry looked off into the middle distance, nursing his drink. The conversation took on the rhythm of small talk, even though the topics loomed large.

'How's your wee brother?'

'Are you close?'

'Are you doing well at uni?'

'Did you have a good life, Kristie?'

I answered, feeling real pressure to lie. Or at least put some shine on things. The truth was not something a father could be proud of. Plus, Barry was still there, still pretending to find the building to his left incredibly interesting.

I asked questions too, and although Shane talked quickly, scrimping on details, I learned that he was long-term unemployed, living in poverty and banned from at least two local pubs due to volatile relationships with the barmaids.

Sip by sip, answer by answer, I tasted disappointment.

Finally grateful for the limitations of the half, I looked at the buildings around us, stone slowly tarnishing in the rain, and blamed the weather for my early departure. As I stood to leave, my mind settled on a word. *Faded*. The prosaic grey of the sky fading into the dull silver of the granite around us, and this man, my father, looking like the rainfall would wash the last of the colour out of him.

Shane offered me the pile of photographs and smiled a little when I accepted. His eyes, I noticed, creased like Matthew's. I put the photos in my pocket and left my hand there, running my fingers over their folded corners and edges while I said goodbye.

Crossing the road further away this time, I pulled my hood around my face, pinching it under my chin to keep out the wind. I heard my mother's voice in my head, 'Well, that was it, that was your father,' and I followed the thought as it echoed in an unanticipated emptiness. Our reunion had not been a return to something lost. There was no thunderbolt of father–daughter connection. Daniel's T-shirt neckline would remain unstretched. I was stupid to think my dad could be the one to save me. He couldn't even save himself.

•

Rich was my Hogmanay kiss from three years before, and the increasing flirtation in our texts had made it clear that we would hook up again tonight. I'd chosen purple skinny jeans, a one-shouldered top in black and electric blue, and my favourite flats that looked like they were fashioned from a vintage carpet sample. I'd straightened my short hair and left it flipped to one side to reveal a freshly shaved undercut. Yes, it was the noughties.

Having dropped Jo off with my mum, I walked to the pub in the wind and rain, wrestling with my giant tartan umbrella. I'd found it in a charity shop and now, I wasn't entirely sure it was waterproof. Feeling my phone vibrate, I extracted it from the back pocket of my already damp jeans. A text from Rich. He would be late, but his friend Janek was already there. He was coming, he promised.

As I contemplated the task of meeting someone new, my pace slowed. I'd only managed a single glass of wine before leaving the house. I was not prepared to be first-impressions-level-charming. Turning onto Lothian Road, I hesitated in the wet pavement glow of a warmly lit bar before pushing open the doors. Leaning my now sodden, heavy umbrella against a wall, I walked to the bar. Damp cuffs stuck like shackles to my wrists, I ordered two shots of tequila.

The barmaid smiled in a friendly but knowing way as she prepared my drinks and I felt compelled to say something. 'First date nerves. Sort of.' I managed a tight laugh as she put the glassware in front of me. Skipping the salt, I swallowed the mouthfuls of room-temperature liquid, winced, then wiggled a lime wedge from between my front teeth. Bolstered by the courage I'd soon feel, I went back out into the rain.

At Cloisters, warm bodies and wet coats made the windows foggy, and I knew the tequila had worked when the hubbub emanating from the door felt inviting. The bar was packed, and I stood on my tiptoes to scan the room. No sign of Richard. Doing one final check, I peered round the corner towards the open fire, and my eyes settled on a man. An absolutely gorgeous man. Leaning on the wooden lintel, his wiry frame was draped in blue jeans and a plaid shirt, rolled up to reveal muscular forearms, unbuttoned to reveal a crisp white tee.

He was tall. He had long hair and a beard, striking in a time before these trends had Edinburgh in a chokehold. Kind eyes. Full lips. I tried not to stare too long. Instead I focused on collapsing the huge fan of my umbrella, doing the maths on how to strike up a conversation that would definitely lead to sex.

'Kristie?'

Elbow deep in soggy tartan billows, I looked up to greet Rich.

But it was not Rich. It was plaid shirt, long hair. The bearded man from my fantasy. Feeling exposed, entirely seen by the universe, I stuttered, my brain unable to grasp what was happening.

'How . . . I mean . . . who . . .'

Smiling, eyebrows set inquisitively, he asked again: 'Kristie?'

I nodded slowly.

'I'm Janek . . .' He paused to give me time to twig, and when I didn't, he continued: 'Rich's friend?'

Reaching out, he took the umbrella from me, grappled with it briefly and slotted it neatly into the stand by the door.

•

My father and I reconnected more than a decade after our first meeting. I was still lost, still struggling, and thought that maybe now he could be something to me. I called him and we talked, bonding over music and our ability to remember any and all lyrics. At one point in the conversation, he asked me, 'So, um, are we human, or are we dancer?' and although I recognised the words from a song by The Killers, I had no response for him.

'Ach, I dunno, Kristie,' he said. 'I'm nervous.'

Shane spoke of his family. I was surprised to learn that he was from a long line of Travellers and of the bigotry they had endured. I discovered that he was a father many times over, including to Anne-Marie, a year or so younger than me, and was in some way estranged from all of his children. He asked about my brother, again.

I'd barely seen Matthew in years. Mum had told me he was working as a bar manager, engaged to some girl. Not my news to share. So I was vague, saying just enough that it felt like I was saying something. Shane described his lifelong battle with alcoholism, which he thought was the cause of large cysts all over his face and head. He suggested meeting up and then took it back. He was concerned. What if his disfigurement got us turned away from the nice places I might want to go?

'Kristie,' he finally said, voice wavering, 'will you forgive me?'

He didn't elaborate, and maybe it was better to allow me to decide what it was that I was forgiving him for. He had neglected and then abandoned me, not once but continually. Never trying to find me, never reaching out, never sending a card or money. He had made a choice not to be my father, every day. I said I forgave him, but my voice didn't waver. This almost-stranger was so full of regret, it was easy to provide absolution.

Maybe that was enough for both of us because we didn't keep in touch.

The call came from Anne-Marie, three years later. Our dad was in the hospital, unconscious and unlikely to wake. If I wanted to say goodbye, I should come now. Even in my shock, I recognised the generosity of her offering – to me, a woman she'd never met. The decision arrived fully solidified in my mind. No. I didn't need to see him. I had grieved already for a relationship that had never existed, and I had the rest of my life to grieve the fact that it didn't.

That night, after the call to say my dad had died, my mum came to my flat with my brother. We drank wine and listened to her stories, now a little more honest. He'd stolen money from her to go drinking, disappearing for days. He'd cheated on her, she'd always known, but the existence of our half-sister confirmed it.

I drank until I could barely string a sentence together, the end of each word dissolving into the next, mourning my father, an alcoholic, by getting shit-faced.

My mum, with hands on her hips, said, 'Gosh, I've never seen you like this before, Kristie.' I hid bitterness behind a smile and a loose shrug. I had been drunk almost every day of the last decade.

With Nell strapped to my chest, her big blue eyes staring up at me from below my chin, I synchronised my wine run with collecting Jo from primary school. Popping a bottle of Rioja in my basket beside Capri-Sun and carrot sticks looked oh so very normal. On returning home, I'd open the wine and drink it while I made snacks and then dinner. By the time Janek finished work a couple of hours later, I had texted him to pick up another bottle. He came home most nights to kids sitting at the table, a beautiful meal and a drunk wife.

At dinner, cutting up little Nell's food, chatting to Jo, I kept track of how much was poured into my glass, versus how much was poured into his. Afterwards, I settled the kids with a storybook in one hand and a glass of wine in the other. All the while keeping one eye on the clock, watching it creep ever closer to its 10 p.m. position. The point at which shops would no longer sell me alcohol, even when asked nicely.

Once the kids were in bed, I'd work on persuading Janek to run to Sainsbury's for more booze. My most charming self would emerge. Slowly shifting the subject to wine, cider, beer. Casually broaching. We could have a nice, relaxing night now that the kids were asleep. My hand on his leg.

Drinking is an integral part of Scottish culture. It is everywhere and attached to everything. When even toddlers' birthday parties are booze soaked, no one interrogates the mother who, after a long day, stays up late drinking £18 bottles of Viognier while listening to Joni Mitchell. No one questions the happy, accommodating drunk whose only sins are to talk too much and insist that everyone be quiet four or five times a night so they can hear the best lyric ever written. My alcohol dependence had been well hidden amid widespread alcohol dependence.

I could see the morning, pink on the other side of my eyelids. Emerging from sleep, my brain was yet to fully acknowledge the sensations of hangover, but I felt them there, lurking. I opened my eyes and the room lurched. My stomach followed. Unsticking my tongue from the roof of my mouth, I tasted vomit.

I turned the handle of the bedroom door as quietly as I could, and stepped into the hall. 'Mumma!' Nell launched herself at my legs, arms outstretched, begging to be picked up. 'Hang on, hang on,' I said, and tried to keep walking, but she was wrapped around my calves. Ignoring her squawks and trying to prise her fingers from the fabric of my jeans, I lost my patience. 'Janek!' Nell, two years old, shocked to hear the sudden snap in my voice, crumpled into tears.

Janek appeared and lifted Nell into the air and away from me. I looked at them, their faces in matching father-daughter disappointment. Then I saw the laptop and empty bottles on the kitchen table. On the worktop, remnants of Jo's birthday cake, carelessly swaddled in wrinkled silver foil.

A thirteenth birthday was special, and I'd done so much planning, starting work on the cake two days in advance. Three tiers of sponge coated in white icing, a fondant rendering of Thor's cape cascading over one side and tiny rice paper comics placed in green icing shaped to make the Hulk's fist. I'd had to stop myself at least twice from giving her a sneak peek.

I searched my memories for Jo's reaction. Obviously, my masterpiece had been eaten. How did it taste? Did she love it? I didn't know. My heart began to beat, strong and irregular. Bright lights danced in front of my eyes. Hands on the edge of a chair to steady myself, I held my breath and waited for the panic to pass. In the hall, I could hear Nell's teetering toddler footsteps as she walked to Jo's room and asked for her sister. Janek explained, exasperation in his voice, that Jo was still at school. I needed them to be quiet so I could think.

Sitting down, I pulled my phone from my pocket. Tapping on the photo gallery, I was relieved to see Jo's smiling face among the thumbnails. Clicking on each photo, I worked my way back through the evening: dancing in the kitchen, cutting the cake, singing 'Happy Birthday', Jo opening her presents, Nell sitting on her lap. Eyes filling with tears, guilt sucking all the air out of the room, I tried to pair each image with an accompanying memory, any memory at all.

•

I was grateful that Rich had been dumb enough to bring a better-looking man with him on a date. Together, though, Janek and I were like chalk and cheese, if chalk and cheese were into emotional bare-knuckle boxing. Both deeply damaged by our childhoods, our shared dysfunction pulled us together. From the start we were each baffled and offended by how the other saw the world.

Soon after we'd met, Janek had returned home to Canada for Christmas. He was a sensitive, adoring *Playgirl* pin-up of a man and I was besotted. But I felt uneasy. It wasn't just the threat of time apart. Here was a solid, structured person, a man of routine and impeccable barbering. Did he really know what he was getting into? Of course he fucking didn't.

I'd always been able to entice the kinks out of men. Even though I knew little of my own needs, I prided myself on being a thoroughly filthy Scottish lass willing to try anything. But Janek was different. A true Canadian, he baulked at my use of the word 'cunt'.

'It's not a shameful thing,' I'd say, teasing.

'No, no, of course, but . . . can you not?' he'd answer, eyebrows lowered.

With two months of separation on the cards, we kept in touch in the most 2009 way: Google Chat. Keen to keep him interested,

I'd asked and asked. What was he into? Anal? Bondage? Orgies? I would almost have preferred rejection over his damning of these options with the text equivalent of a shrug. When he'd suggested we share porn links, I knew I'd won. This was his way of finally revealing the currents of deviance that lay beneath a placid, polite surface.

Scrolling through Pornhub, I tried to find a video that I had recently watched (and enjoyed). On the rare occasions I watched porn, it was usually the same sort of thing. A threesome, two women and a man. An absence of fake boobs and fake orgasms was a bonus. Real, wet, messy fucking.

I scrolled through thumbnails. Titles like BIG TIT SLUTS CUM UNCONTROLLABLY and HOT THREEWAY WITH BIG DICK STRAP-ON DESTROYS HER rolled by until I found the video. TWO CUM SLUTS FACE FUCKED HARD SLOPPY SQUIRTING THREESOME. Well. I hadn't remembered that as the title, but there it was. What he came up with would be worse.

Nonetheless, pasting the link into the chat field, I cringed. I waited.

A message appeared. 'On three send links?'

He was typing again. Then the word 'one' appeared on the screen, then 'two', then 'three'.

I hit send on 'three' and our links came through together. His first. Mother Teresa looking down from the highest horse.

PETITE BRUNETTE GIVES AWESOME HANDJOB.

•

Rough or gentle, messy or neat, Janek and I agreed on nothing, and over the years this added layers of misunderstandings and resentment. Then Nell was born, and the stakes got higher. It didn't matter what it was, when I asked for help, he always had a reason that he shouldn't, or couldn't. When I asked him to feed and settle Nell overnight, he looked over his laptop screen in irritation and told me that it was 'impossible'. Furious and exhausted, baby in hand, I told him that he clearly didn't understand what that word meant.

Nell was nine months old when postnatal depression kicked (my face) in. I'd hear the tempo of Jo's footsteps approaching my closed bedroom door and feel panic rising. What did she want? What did she *need*? When I breastfed Nell, my very essence was being drained from my body into hers. I sank into bed, depleted, and there I spent most of my days, rolling away from Janek in the nights. My skin raw from being touched, even gently.

•

Unable to deal with the stress of being seen, I had waited until Janek took the kids to the park. Getting out of bed and opening the door, I lowered my head and listened, wincing. Even silence felt oppressive. In the bathroom, I tripped over a plastic pirate ship, too cheery on dark tiles. I didn't have the energy to swear and continued through the water next to it with my pinkie toe throbbing.

Hunched over on the toilet, I realised that I stank. Weeks' old sweat and the fainter metallic scent of my period blood filled my nostrils. I looked at the tiny glass cubicle next to me.

Get a towel; pick up the funnels, Tupperware and whisks from the shower tray; turn the shower on, remove my clothes, get into the shower; get wet, wash hair, wash body, get out, dry off, get dressed.

I didn't have the headspace for even half of it.

Once I finished urinating, I began to change my pad. These days everything took so long. Basic tasks were punctuated with sigh-filled pauses as I waited for my brain to find me in the fog. Careful not to do anything too quickly or it might lose the signal altogether.

Pad stuck lopsidedly to my underpants, I waited with elbows

on knees and head in hands to find the wherewithal. Looking down the length of the bathroom into the hallway, I saw the evening light, filtered through leaves, dancing on the walls. I remembered a time that this sight would have brought me joy.

But now, I felt furious. I'd told him that to recover from PND I couldn't feel needed, that any obligation might push me completely over the edge. My eyes moved between the busted basket that held hats and scarves, scuffs from Jo's scooter on the laminate (how many times had I told her not to?), and two bags for the charity shop that had been there for months, one now spilling onto the floor. Janek was useless at staying on top of it all.

Then I saw the marks on the wall by my bedroom door. They must be shadows, part of this evening's light display. Maybe. I stood, pulling up my stained pyjama bottoms with their spent elastic waist, and walked into the hall. No, the marks were real. I felt a terrible heaviness before my brain informed me of their origins. Maybe it was that slowness I had been experiencing or maybe it was a kindness keeping the truth from me a second longer.

On the white paint, two lines of soft grey finger smudges. One at Nell's height, one at Jo's, where hesitant little hands had paused at my bedroom door many, many times. Unsure, fingers lingering on the wall, they'd listened and considered, before ultimately walking away.

I touched the marks, tracing my daughters' fingerprints. A memorial to my inadequacy as a mother. Heart leaden, shimmering golden hour bokeh dancing all around me, I knew that my children would be better off if I were dead.

•

I was desperate enough to take arsenic but Svea suggested Bupropion. It had been, my friend said, quite helpful in not killing herself. The doctor said no. Bupropion is available on the NHS for help with smoking cessation, not depression. I told him I'd always wanted to take up smoking. He didn't smile. Would I try counselling instead? With a new baby, I was high priority: I could join the ten-month waiting list.

I bought the pills from an online pharmacy and had them sent from America where Bupropion is widely prescribed (for depression). I messaged Svea when they arrived. 'Good luck,' she typed into the chat field. Within weeks, my bedroom door had opened, my daughters once again climbing into bed beside me.

The relief of emerging from depression was tempered by the presence of its spectre. The looming bodach of worry and recollection. It was hard to forget the months I had spent telling myself that I was worthless and working up the energy to die. Warily, wearily, I tried to pick every piece of me up off the floor and assemble who I was from memory. Sorting through the debris, not only did it seem like essential parts were missing, but some, long forgotten, had reappeared. One of those was a mixed-race woman.

I'd been raised in the aftermath of extreme racial violence: the abuse my mum and her family had endured in Glasgow. I'd endured racism in school and beyond. I had never identified as white, but it was confusing. Often, the strongest connection I had to my racial identity was being treated badly because of it.

And every year since leaving Caithness, I had leaned more into my light skin. It was easier. I hadn't had the capacity to present, and defend, that other aspect of myself to the world. Maybe I'd given up my right to identify as mixed-race.

So I began to read about it. Articles, books, posts on social media. Experiences of being othered, objectified, shamed and fetishised. As the term 'abuse' had shaken my foundations a decade before, I began to identify 'racism' as the name for so many of my worst experiences. But if I had experienced racism, I had also experienced light-skinned privilege.

I smelled it at the bus stop. Rose petals, powdery cotton and warm female skin. Cutting through traffic fumes, it transported me to a softly lit courtyard on a summer night. As the number 7 bus pulled up, I was brought back to the reality of a cold, wet Leith Walk and people jostling, elbows out, to board first.

Now at the back of the bus, I unhooked the strap of my bag from across my body, placed it on my lap, then put in my headphones. As we pulled away from the kerb, a brown woman, hands laden with shopping bags, hauled herself into the seat opposite me. Placing her bags beside her, she settled the fabric of her coat.

I recognised the heady scent from the bus stop and began to watch her out of the corner of my eye. She reminded me of my mum and my aunty. Long dark hair, perfectly plucked eyebrows and, though they rested quietly in a way my mother's never did, silver bangles on her wrists.

I took out my headphones. 'Excuse me?'

She smiled and said nothing. I realised she was waiting for me to continue.

'I was just wondering what perfume you're wearing. It's incredible.'

She held up her finger, then turned to rummage in her leather handbag, reappearing with a monogrammed black satin pouch. Loosening the gold drawstring, she extracted a purple bottle with gold coloured hardware.

'It's Aoud Purple Rose . . .' she said, handing it to me. 'Not cheap.' Her eyebrows rose to emphasise the point.

She rolled her Rs. I wondered if her accent was west coast, if she was from Glasgow.

I lifted the bottle slightly in acknowledgement as I handed it back to her. 'Thanks for this,' I said and she slipped it back into its silky purse. I unlocked my phone to write down its name. Typing, I asked, 'Where are you from?'

I looked up, expecting her answer, but instead I saw a tense face and dark eyes examining me. When I realised what was happening, my stomach fell. What kind of arsehole did she think I was? I grasped at an explanation, but the words slipped through my fingers.

'Oh. No, that's not . . . I meant . . . My mum's from Glasgow . . . I'm . . .'

But it was too late. Legs leaning away from me, torso angled towards the aisle, head down, she kept her eyes busy with her phone. The damage was done.

My thoughts raced as I sat in shame. How could she lump me in with those people who actually ask brown people where they're from? This was different. I was different. I was mixed-race, I had experienced racism, I had been asked where *I* was from! She couldn't see any of that. And apparently I couldn't see past my own experience, hadn't been asked that question enough times to know not to ask it.

To her, I was the kind of arsehole who refused to believe she was Scottish, who was determined to expose that, at some time in her life, she or her ancestors belonged somewhere else.

Seeing my stop up ahead, I stood to leave. I wanted so badly to say something, to fix it, for her to understand. But I had already tried, and this was my burden, not hers. As I passed, my bag brushing against her knees, all that came out was a mumbled 'I'm sorry'.

•

Nana's house was full of the forced cheer of people we barely knew. My mum was in the kitchen beside Uncle Anwar, the brother who had punched her. With Nell in my arms, Janek and Jo at my side, I watched through the door as Mum fussed with a spread of sandwiches, biscuits and cakes.

When we walked in, she barely lifted her eyes from the circle of miniature Battenbergs on the plate below. As if greeting me would upset some meticulous balance. Maggie, my uncle's wife, generously broke the silence.

'It's good to see you, Wee Hen!' Like she had seen me last week, not ten years before. She nodded at Jo and Janek, then reached over to stroke Nell's cheek with the back of her index finger, long red nails conscientiously pointing away. Nell sneezed.

This house was newer, smaller, than the one I used to visit and had no room for Nana's collections of teapots and ceramic cottages. The antique fire guard stood useless, protecting us from the unpredictable threats of a beige wall. Nana was in bed in a room that smelled faintly of baby powder and disinfectant. When Aunty Jane saw Nell and me at the bedroom door, she stood and switched places with me. It was my turn.

Opposite Nana's bed, a wall hung with family photos. Even those long estranged could be seen in portraits with 1990s haircuts and shoulder pads over thin brown arms. Cousins posing in T-shirts with the names of TV shows long past. The date printed in bright yellow lines in the bottom right corner. Nana, soft and frail in her bed, moved her eyes, again and again, between the faces on the wall.

Then she saw Nell in my arms and perked up. 'She wants great-granny, doesn't she?'

I agreed that she did and settled Nell on the bed, pretending not to hover. Nell, babbling and drooling, received generous compliments on her storytelling skills, but after a few minutes pushed out her bottom lip. I lifted her back onto my knee so she wouldn't cry.

In the hall, my aunty Maggie was wiping down the door frames with red hands and a cloth that was so hot it was steaming. My eyes filled with tears.

'No, no, no,' she said, bringing her eyebrows together as she reached out to awkwardly pat my shoulder with a hand that smelled antiseptic. 'Don't do that!' And then, more quietly, eyes wide with meaning. 'You'll upset her.'

She wasn't trying to be mean, but she had to be firm. I understood. I shouldn't cry because no one had spoken to my Nana about her dying. And they weren't going to.

In the living room, still holding Nell, I gingerly moved aside my mum's coat and sat down on the small sofa.

Uncle Anwar was asking Janek if they had Mr Kipling in Canada; my mum and her sister Jane were in the kitchen making more sandwiches although the table was still covered in uneaten sandwiches. As the only kid over the age of one, Jo, also in the kitchen, was being accosted by every passing relative. Did she know how to knit yet, did she have a boyfriend, had she tried the peapods in the garden, did she want to sing or do a dance, did she know the song from *Joseph and the Amazing Technicolor Dreamcoat*?

No one mentioned why we were here together for the first time in fifteen years. No one spoke of Sajjad or Clare or their long absence. It had always been like this. Attempts to talk about the past, or feelings, punished with silence; the loudmouth, the troublemaker being pushed out.

Because an acknowledgement of anything was an acknowledgement of everything.

So we had entered an experimental theatre performance. I was a member of the audience but also a member of the cast. I had a responsibility to perform. As the least practised cast members, it would inevitably be me, or Jo, who broke, and then my aunty Jane would tut in disapproval or maybe she would leave the room in annoyance, and my mum would tell her not to dare talk to her kids that way and everyone would yell and scream and that would be my Nana's last memory.

I'd known they were fucked up, that they struggled to express themselves, but I'd always assumed that in a big moment, a moment that really mattered, a moment of definitive no-take-backsies, they would reveal their humanity. But here we were, on stage, and nobody was allowed to grieve.

I couldn't do this. I wouldn't make Jo do this.

When I went back into Nana's room, her eyes were closed. She heard me, lifted her arm, and patted the bed to show she wanted my hand. The slightly tensed muscles in her face told me that, despite a cocktail of painkillers, she was still in pain. I watched her try, several times, to open her eyes.

'Just rest,' I said and, choking back the forbidden swelling in my throat, I kissed her cheek. She must have felt my tears, wet, or the tremors of my body, because she said, voice distant, 'Oh, don't cry, Kristie. I'll see you soon.' Nana was on stage too. The leading lady.

Moving through the house, collecting our belongings, I said friendly, probably manic, goodbyes. Three generations' worth of pain was stored in the cells of my body. It lived within me. I had been built around it. I was overfilled, about to tip all over the floor, and here was my family, unspillable. I needed to get out of there.

In the car on the way home, I faced forward in the front passenger seat, numb. Janek focused intently on the dark roads, kids worn out and quiet in the back. I measured my breaths, thoughts careening through memories. It took an hour on the road before they collided head-on with regret.

Bye, Nana, love you. I had said it like I was going to see her next week. Rather than telling her, really telling her, how much I loved her. I knew so little about her, I should have asked. Her childhood, her life, what my mum was like when she was little. At the very least, how much I loved her.

I turned on the radio to stop myself thinking about how Nana must be feeling. 'This Is What We Came For' played softly from the speakers near my feet. I was sure she had some idea that this was the end. Going over her life, reliving it all. The good and the shitload of bad. Fighting her own grief, understanding she would never see her children Sajjad or Clare again.

Quietly, so as not to wake the girls, I began to cry.

My mum called two days later. And as she told me, flatly, that Nana had passed away with three of her children stoic at her bedside, I knew that I didn't want to perform on stage ever again.

I couldn't hide behind my light skin any longer.

•

After nearly ten years, I had accumulated a large Edinburgh peer group. Good friends, all of whom were, on paper, liberal. 'On paper', not in a disparaging way, but in that difference between believing something should be the case, and how you act when you find yourself in that situation. You know what I mean.

One July afternoon, sitting with Janek and our friend Stuart at our cheap pine kitchen table, drinking and chatting, the conversation turned to the recent Brexit vote.

Stuart, blunt as always, said, 'We need to end the union. England is a racist little shithole.'

I shrugged. 'But it's always been that way. Now it's just closer to the surface. In Scotland, too.'

Stuart looked surprised. 'Racism? I don't think so.'

Taking a large gulp of wine, I braced myself before saying, 'Well, as a mixed-race woman, I see it, especially the casual stuff.'

There was a pause. Stuart raised his eyebrows very slightly, saying, 'A mixed-race woman? How . . .?'

My heart beat faster. I told him, carefully, again. 'My grand-dad was Pakistani, my mum is mixed-race and . . . I'm mixed-race.'

For a split second, Stuart looked uncomfortable, but then he was laughing. 'Is *that* why you smell of curry?'

I hated myself for not expecting this. The shame of the Thurso tuck shop came flooding back like it had never left. My face ignited with a heat I prayed was noticeable only to me. And then I laughed too, forcing myself to make it sound as real as possible. I stood up from the table, walked to the sink and filled a glass at the tap. Just like that, it was over.

With Louise, I tried a different approach. As we sat in her living room, again drinking wine, I tentatively explained the journey I had been on. I was nervous, and when I'm nervous I over explain.

'I always felt mixed-race, but it was sometimes easier to let people think I was white, you know? My granddad is Pakistani, even though you maybe can't see it in me, and it's not like that matters because Pakistan is actually a pretty diverse place. In the north, where my granddad is from, there's a lot of light skin, even red hair and blue eyes in some places. I've been reading about it. I'd like to go someday, I have relatives there, but I just don't have the family last name because my mum changed it. Did you know my brother's middle name is Faqir?'

Finally, I paused for breath.

Louise nodded and looked at me with sympathetic eyes over a long sip, burgundy liquid in her mouth preventing an immediate response.

Unable to gauge her reaction, I continued. 'I've wanted to say this for a long time. I suppose I wasn't brave enough–'

She brought the glass down and held it in both hands, but her face still looked a little blank.

'–that I'm mixed-race.'

Smiling in understanding, she replied, 'That's great. If you want to say you're mixed-race, say you're mixed-race.' Relief didn't have a chance to fully wash over my body before she added, 'But I'll always see you as a white woman.'

It felt like I had been slapped. My face tightened against my skull as I fought to put on a smile. Louise went on talking about other things, but I was lost in thought, trying to quantify exactly what she meant. The part of me that wanted to ask was sternly talked down by the part that understood that the answer might hurt even more.

After the failed attempts to explain myself in person, I began to write. I wrote about my past, my feelings, the ways that racism followed me. I wrote about body hair and self-acceptance and being mixed-race in Scotland. Messy and verbose, these unburdenings should perhaps never have found their way onto the internet. Looking back, I see that I was desperate for validation and acceptance. But no one cheered my honesty or wrote messages thanking me for making them aware of how racist they were. Instead, strangers told me that if I hated Scotland so much, maybe I should go back to where I came from.

On my private social media, I was pretty sure I noticed fewer likes on my photographs, dwindling invites and a strange formality to conversations. As if, all of a sudden, we weren't people who had seen each other vomit. I was ready to transition from the titbit 'My granddad was from Pakistan' to a full-hearted declaration of being mixed-race. But maybe my corner of progressive Scotland wasn't ready to hear it.

After months of convincing myself that I was imagining things, I looked at my accumulated data points and told myself the truth. Things had changed. Seeing me as white, my friends had treated me as one of the gang, someone they could trust to hear their good intentions rather than their words. And there I was, being not white, changing the parameters.

•

There was deep dysfunction in our relationship, but Janek and I changed each other's lives. I supported him through an Autism Spectrum Disorder (ASD) diagnosis, recognising the pain and emotional neglect of his childhood, and a long overdue haircut. In return, he financially supported our family, saw me through severe depression, getting medicated, and the grief I felt about his haircut. He also encouraged me to pursue a career in photography.

In my teens, the art department had felt safe. I'd spent my

lunchtimes and free periods working in spaces filled with the smell of drying paint and murmurs of 'thank you' or 'excuse me' from other kids seeking sanctuary. Hours in the darkroom, the soothing red light, slick sheets of wet paper, one eye on the timer. I really would lose myself in it.

Art school was for rich white girls who wore crushed velvet bucket hats and platform boots, and I was neither, owned no velvet. So when I'd left high school, I'd left photography behind too.

My return was through the iPhone 3, its two megapixel camera. I snapped the important things in my life. Jo's first day of school, her face serene as she slept on my arm, and many, many close-ups of flowers. Slowly, months of casual documentation shifted to something more meaningful.

Walking along the canal, my eyes drawn to an errant patch of light, I felt fortunate to see it just so. Taking a new route to town, the disparate lines of the city converging in visual balance, I wondered if I were the first to notice. I started to itch, craving to capture everything I saw, every interface between myself and the universe. My equipment was humble, but with it the world's overlooked beauty felt relevant, and, in a small way, I did, too.

When Janek offered to buy me my first 'proper' camera, I was apprehensive. My interests tended to wax and wane, and, truthfully, I had no idea how long I'd be into photography. I didn't want to disappoint him or waste his money. All I could say was, that right now, I loved it. He told me it was enough.

•

In Canada, Janek had been Mr Outdoors. He'd go camping in Algonquin, wild swimming, kayaking and then portaging for miles. Most of his Facebook photos had been taken around a forest campfire, from the shore of a lake or at a climbing wall. He had been trying to persuade me to go outside for years.

To visit this thing on that hill, to hike between those towns and these villages, or 'maybe a short walk to the shore?' Every time I'd said no.

I wasn't merely disinterested. I hated it all. More than a decade after leaving Caithness, I hadn't set foot anywhere more wild than Edinburgh's cycle paths, and the obligatory one-time walk to St Anthony's Chapel and Salisbury Crags just to say we'd done it, which doesn't count. I shook off jolly autumn leaves hitchhiking on my shoes as if they were dangerous insects, sat fully clothed, boots tightly laced, as the children played on the beach.

I didn't even want to have good memories of the outdoors.

My new camera had shifted this outlook, and over time Edinburgh's landscapes had begun to feel uninspiring. My desire to photograph everything had me inching further and further towards the city limits. Finally, out of options, I agreed to a New Year's Day walk along Harlaw reservoir, if Janek absolutely promised not to get his hopes up.

At the water's edge, no longer sheltered by trees, the wind stung my cheeks. I lowered my camera, cupped my hands against my mouth, and blew hot breath onto tingling fingers. The surface of the reservoir was an extensive, icy archipelago, each fragile floe shifting in the wind. With movement so smoothly regimented it could have been mechanical, each little island was making its way to shore before breaking and piling onto the patient pebbles below.

I pulled the collar of my waterproof over my nose and tried to ignore the condensation that formed almost instantly on my upper lip. Behind me, I could hear Jo's whoops of joy as she smashed ice sheets with exaggerated karate chops and spinning kicks. Moving away from the water towards the woods, I saw low light filtered through trees, warm spotlights on a world of blue and grey. Camera pressed to my face, I committed these details to its memory, and to my own.

The faint scent of recent rain and leaf litter and, then, a

familiar soft resistance under my right foot. Mud and moss, half frozen, flexing under the weight of my body in the most satisfying way. Bringing both feet together, I jumped and landed in the memory of childhood winters. I saw my brother beside me playing on the partially thawed scrubby land around our home in Caithness. We had come up with the term 'chewy ground', and, walking back down the track, had jostled and shoved, each insisting they had been the first one to say it.

After Harlaw, the trips came thick and fast. Most of my memories of that time are seen through the viewfinder of my camera. Jo running through Glen Quaich, arms held wide and face turned towards the sun; the last rays of winter light on the Tarmachan ridge, earnest against a dusty blue sky. Dark hills, steady gatekeepers to the long and lonely road through Glen Lyon. Nell playing with pebbles at a table, a subtle watercolour of hills stretching for miles behind her.

•

My mum was on the other end of the line, telling me I had to take a new probiotic. I ignored her and tried to change the subject to something less controversial (Scottish politics). Upset that I wasn't following her advice, she lamented, as she always did, that her 'wee girl' had 'disappeared aged eight and never come back'. I sighed. Last week she'd told me it had happened at age ten, and a month ago, she'd said thirteen.

For the longest time I hadn't really known what it meant. I'd guessed it was nostalgia, something that all mothers felt. Now, a mother myself, I knew that wasn't the case.

The story was that, at some point, the 'real' me (sweet, compliant Kristie) had gone away and was replaced by sick Kristie, or bad Kristie, or broken Kristie. A changeling who would not go gently. So my mum, shaking her fist at the sky, was left valiantly trying to figure out what was wrong with me.

To vanquish the monster who had stolen the daughter she loved.

'I sent you a link,' she said.

I sighed again, louder. But I opened my laptop and looked through my overflowing inbox until I saw her email, subject line all caps and exclamation marks. It led me to a website taken straight from the internet archives. Lots of yellow, ugly fonts. Essentially a giant chat room. A never-ending column of text from people whose thyroids weren't working. Ah. Hypothyroidism. The newest arrival in her long line of amateur diagnoses.

Immediately overwhelmed, I said that I didn't have time, and lied that I'd read it later.

That evening, in bed, back online, the tab was still open. One yellow word caught my eye. *Trauma.* I clicked through to an article on how adverse early childhood experiences were linked to autoimmune diseases. And at the bottom, a checklist of hypothyroid symptoms. I ticked 'yes' on all of them. A revelation.

The NHS deals with women's health issues, including hypothyroidism, like the Tories deal with poverty: it's all in your head and maybe you could try not complaining so much? Even after an endocrinologist confirmed the condition, I struggled to get the right treatment.

My mum offered to pay privately for the medication I needed, and although I didn't completely trust her motives, I gratefully accepted. It was the perfect compromise. Sure, there was something wrong with me, physically, in my body. But my mum, and how she had treated me, had caused it.

At the playground a few months later, Nell on the swing, Mum had told me between pushes that she'd noticed I'd been calling her a lot less. She'd said it was a sign that I was happier, healthier, more resilient. I listened, halfway between amusement and a never-ending internal scream, as she said that we had finally fixed what was wrong with me.

When I started calling my mum to rant about racism, I think she worried that I was unwell again.

Since Nana had died, I'd thought a lot about the past, and my family. The lustre of childhood stories had worn away, and for the first time I could see this history for what it really was.

Children's feet pounding dark city streets. A broom handle, a plastic baseball bat, a plank of wood pulled from an old fence clutched tight beneath small, pale knuckles. I considered the dread coursing through my mum's eight-year-old body, afraid she and her siblings would be killed or maimed. All three of her brothers were stabbed. Blood, pavement, torn clothes. And, once those long nights were over, each of them forced to return to that same world to fight again for a place within it.

It was disorienting to realise what my mum had been through. She was so strong and so fierce, and I was afraid of her, but she had once been vulnerable, the victim. I thought of my Nana, too. How she had made up her own mind about who to marry, worn a gold pendant of her name in Urdu on a chain around her neck, believed in a brave new world where mixed-race families thrived – and then watched her dream ripped apart.

They'd had no way of protecting themselves on the outside, so they'd survived by walling themselves off, putting spikes on top of those walls, and trimming them with barbed wire. Even though it was now clear to me that racism was at the core of the stories I'd heard as a child, it had never been directly mentioned. By anyone in my family. And even though I'd been called a 'Paki' nearly every day at school, my mum and I had never talked about it.

'Mum,' I said, 'did you see what I posted about racism in Scotland?'

Once my sounding off had subsided, and I had convinced my mum that my thyroid meds were still working, it felt incredible to talk about these things in extended late-night phone conversations. To hear her voice light up in recognition, growing stronger, as I had, by learning a language that described what she had been through. Still went through.

For my mum and her brown skin, racism, its limits and its scars, were inevitable. The shop assistants whose eyes narrowed as she approached, their tone changing, curt and unhelpful. An evening among new friends as one slowly revealed themselves to be someone who used racial slurs, while others revealed themselves to be people who laughed nervously when someone used racial slurs. Then there were her own family members.

With no solid information, only an intuition that things needed to change, she'd tried to get her family to open up. But it had never worked. Her attempts would end in terrible fights. Shouting, cruelly personal. Fingers pointed, usually at her.

Her family could never acknowledge that racism had played any role in their unusually difficult lives. It would be too painful, she said, after everything they had been through, to accept that Scotland still didn't see them as definitively Scottish. She told me how hard it had been for her to be pushed further and further away, scapegoated as the one causing the problems when she was the one trying to fix them.

I told her about being asked, so often, where I was from. And how, despite that, I had enacted the same microaggression against the pretty perfume lady on the bus. When she laughed softly at my stupidity, I felt a weight lift off my shoulders.

It's hard to explain to a white person the significance of microaggressions, or sometimes even what a microaggression is, but none of this was necessary. My mum understood exactly. I'd spent a long time out in the world looking for belonging, assuming it lay in some unknown place. But here, with my mum on the phone, it finally felt like I existed in the world alongside someone who really understood.

Acknowledging one thing meant acknowledging everything.

•

In British society, we routinely (and conveniently) lay the crimes of violence at anger's feet. We ignore its benefits, and see even righteous anger as dangerous. So absolute is our distaste that even speaking directly is unacceptable. And then we get drunk and fight in the streets.

As the generations of pain worked their way to the surface of my body and my understanding, I noticed that the higher it rose, the more it looked like anger. And I welcomed it.

Although anger had been pitched to me as irrational, something that distorts and deforms, my internal experience had been different. Through anger I could access parts of my brain that were normally off limits. My mind became sharper, my tongue too. No one had ever been so convincing about the dangers of nuclear holocaust as I was in Miss Hewitt's second-year English class; I was a rhetorician, a top lawyer skilled in savage cross-examination. Erin Brockovich in the scene where she says, 'That's all you got, lady. Two wrong feet and fucking ugly shoes.'

In pain, I felt stuck. All I wanted was for the person who hurt me to fix it, and I lived in hope, waiting for that day. Pain kept me trapped in impossible longing, but anger led me towards action. It didn't feel like it was festering away inside me, compromising my very soul. Anger felt vital and energetic. Like, if I performed a little alchemy, it was all the change I could ever need.

•

We sat outside at the café beside the tree-lined cycle path, Nell squirming on my knee. Steam from Jo's hot chocolate rose in a steady plume from the middle of the table where I'd moved it, away from the exploring, grabby hands of a bored toddler. Jo was telling us about her day at school.

I saw her building up to it, nervous eyes, front teeth biting her bottom lip.

'What actually is a Paki?'

'Oh!' I blurted out, too loud. Then silence as my mind scrabbled for the answer. She looked at me, expectant.

I let the mother I had created, the patient and progressive mother from books and studies and forums, explain that it wasn't an abbreviation of the word 'Pakistani' but a racial slur. Not a kind thing to say, and not a word we use in our family. Tears in her eyes, Jo wanted to know why someone would say such a thing to her, to anyone, and as I told her that it was truly beyond my understanding, my heart resurfaced, buoyed by a current of fast-moving anger.

Jo saw herself as white, but the kids who called her 'Dirty Paki' strongly disagreed. I remembered my mother's admonition. *Never, ever let them see you cry.* Now I understood both the love and the outrage behind those words.

Feeding Nell bite-sized pieces torn from the muffin in my hands, I thought of how I'd allowed friends in high school to call me 'Paki'. A reward for their friendship. For speaking to me. I'd felt unable to ask them to stop in case what was currently a 'term of endearment' returned to being a deliberate racial slur. No matter how much I had wanted it to mean something else, with muscles tensed, stomach knotted, vision and hearing sharpened, my body had known better.

As the weeks wore on, Jo continued to return home from school distressed. As she told me about the continued comments, always from one group of boys, it became clear that this was a bigger issue than I had wanted to believe. I called the school and was immediately transferred to a senior staff member. They said all the right things, and, in a serious voice, promised me they had a zero-tolerance approach to any forms of racism. I told myself to relax, that things were different now. I could keep my anger on a low simmer.

Then it happened again. And again. And again. After many phone calls, many times of hearing the same empty promise in the same overly sincere voice, I asked for an in-person meeting.

Miss Kelcie leaned forward, body language all business. 'The children who were involved have been spoken to,' she told me, 'and I can assure you they had no idea of what that word really meant.'

I smiled, the space between my eyebrows creasing slightly. I enunciated my Ts. 'I'm so sorry. I'm not sure exactly what you are telling me.'

'While we don't condone the use of the word, intention must be taken into consideration.'

Her tone had a sense of finality.

Anger jumped from my stomach to my chest as I said, 'Just to be clear, because being clear is important, the word you are referring to is "Paki"?'

'Yes.' She nodded slowly, a pained expression on her face.

My brow furrowed further, and I felt my body moving with each thumping heartbeat. My voice was unsteady now. 'But this has been happening for months. I'm here because nothing has been done to make it stop. It *has* to stop.'

She sighed and said carefully, 'While I appreciate that Joanna is upset, there is a certain amount of learning that must be done on how to cope with this sort of thing.'

I spoke carefully, too. 'A certain amount of learning to cope with . . . *racism?*'

Her response was too quick. 'I'd appreciate it if you didn't put words in my mouth, Mrs De Garis. I've already explained that the intention of the students wasn't to be racist. When it happens, the school takes racism very seriously.'

Jo was being called the names that I had been called twenty years ago. The same names her grandmother had been called twenty years before me, and the same names her great-grandfather had been called for decades before that.

When it happens. Very seriously.

•

Returning from our trips away, the city often felt inflexible and domineering. Buildings blocked views, streetlights won the sky: it was a tight, tight fit. And our wee tenement flat, lovely, but a space we had outgrown. When we went to look at bigger apartments, we wondered how we could ever afford them.

A decade before, a life in Edinburgh had assured me of many things: acceptance, diversity, cultural experiences, a flat with enough bedrooms. It had given us many things but hadn't exactly delivered on its promises.

Speaking to friends about leaving, we were exposed to a particularly contagious form of FOMO.

'My sister and her husband left the city for some tiny village. They came back after a year, said it was awful, nothing but mud and no buses.'

'I'd be worried that everyone would stop inviting me to stuff.'

'But what will you *do*?'

Shit. What *would* we do? I didn't want to become irrelevant while also dying of boredom.

In the weighing of our choices, we planned a road trip.

Years on, those three weeks with Janek and the kids are still preserved as a set of perfect picture-postcard memories in my mind. A black and white collie leaning lopsidedly against the rusted red body of the Kylerhea ferry. A windy picnic, such a civilised thing, among the wild architecture of the Quiraing. The otherworldly landscapes of Lewis and the silence of the view from Achmore to Loch Langabhat. Machair and mountains in the south of Harris, the soft moonscape of North Uist. The Summer Isles, seen from Ardmair, dusky cut-out silhouettes against an almost violet evening sky. Ben Hope resplendent under a heavy golden sun, flying insects glittering in the air.

I shot thousands of images and discovered that some landscapes generously and faithfully translate themselves to photographs. All the landscapes I mentioned above fall into that category. The geographic photogenic. Then there are the places that remain elusive. No matter how hard I tried, those (mostly northern) landscapes would not be conveyed and their scale, textures, colours, atmospheres roamed free, remaining stubbornly uncaptured.

As a child, I had actually experienced the outside world. Wordlessly, unmediated. Out in the landscape, fully open to all that was around me, I had created memories that I could not only see but feel. Now, although photography had offered a way for me to relate to the natural world, it was also acting as a buffer, keeping full connection just out of reach. For a while, I had needed that.

But in the wild, uncapturable places, it shifted. The unseen stepped forward to meet me once again. On that trip, I accepted that the only way to truly record these landscapes was to experience them.

•

We all have that cupboard (Janek and I had several) piled high, jammed with stuff. A dangerous Jenga tower of overflowing bags for life, bed linen and cleaning implements pressed against a door that needs your full body weight and a couple of firm bum bumps to close properly.

Likewise, we all know that moving house gives you an opportunity to declutter, and I was determined to be brutal. If it had been in a cupboard for five years, we didn't need it. I pulled toys, unworn shoes and a fondue pot out of the dark, shambolic space. Underneath it all, a forgotten bag; hardy survivor of previous house moves. Heaving it from its place between my nana's old dolls house and a clothes horse, I untied the knots in its handles for the first time in years. There were old clothes – mine, which I bravely put in the bin, and some of Jo's, which I added sheepishly to the ever-growing 'keep' pile.

At the bottom, my fingers touched an item wrapped in

a bag. Even before I removed the plastic and felt the soft, black fabric in my hands, I knew what it was. I carefully unfolded the dark cloth and saw the stain of his semen, a chalky bloom. Still visible all these years later.

I remembered. My desperate apologies to Luke and my mum, the bruises, cuts and Monopoly money. The morning-after pill, shame-filled trips to the sexual health clinic. How I'd bragged to friends, pretending to gloat about the incredible rough fuck on the floor of the bar where he worked, needing to be any woman but the one I was. Texting him three days afterwards.

How I'd left Jo.

Hands shaking, I asked myself why I'd kept the dress. It felt like a really, really fucked-up thing to do. I tried to breathe while a mass of emotion twisted in my chest. It hurt. It hurt so much.

And then I knew. I'd kept the dress I had been raped in because it was the dress I had been raped in and someday I would be ready to acknowledge that I had been raped.

With my back against the door, I held it in my hands for a long time.

Laying the Foundations

Seeking an anchor point, I thrust one end of the pinch bar under the boulder. Nothing holds. I take my hands off the cold, rusting metal and rub them together. They consider coming back to life. Standing strong, I shove the bar firmly in place and kick a smaller stone into position near the end of its length. It will be the fulcrum for my lever.

Typically, the largest stones in a drystone wall are found in its foundations, and depending on the project, they can weigh hundreds of kilograms. I push down but the boulder doesn't move. Eventually, all my weight on the metal, feet sliding in the mud, it shifts a few inches and the process begins again. Even with this age-old method of moving stupidly heavy objects, progress is slow.

It's important to be reminded of life without convenience. It's unhurried, often infuriating. It's demanding, and there's no way to the other side but through. You can complain as much as you want (and on a cold, rainy day, I do), but the stones won't move any faster. They exist on their own timelines.

To build with stone takes you back to the fundamentals, and over time it changes your mental pace. It helps you to understand that sometimes there are no shortcuts, that there is no point in focusing on the goal when you must focus on the task. Drystone works the impatience out of you.

We lived our first summer in Perthshire quietly. Enjoying the luxury of a garden, we sprawled across the grass, sitting up to speak to curious neighbours who appeared, now and again, at our front gate. The couple with a daughter the same age as Nell, the red-haired woman with a bike; the man with a cane who walked beside an ambling overweight Labrador. Children came to our door selling homemade jam and we watched them leave victorious, small jars rattling in their wooden cart, coins safe in closed fists. I cut flowers, arranged them in vases throughout our home, and slowly got to grips with who was who in the village.

I had it on good authority that, in a pinch, Albert up the road was a decent chiropractor. The woman in the post office, pointing to a faded photo behind tatty plastic, told me that she had once served Gerard Butler and thought he was lovely. One Friday, out for an evening walk, I met a drunk man on his way home from the pub. Lighting his cigarette after the third try, he swayed as he told me about the time he fought off several much larger men and was offered a 'gypsy king's daughter' as a prize. Eyes misty with drink, he leaned closer, looked me up and down and asked, '*You're not a gypsy, are you?*' He flicked his cigarette in the direction of a lamppost, shrugged, then staggered away.

•

The email from a well-known Scottish radio show took me by surprise. My online writing had, apparently, been garnering some attention (beyond the trolls), and the host wondered if I would take part in a discussion about racism in Scotland. Once I had deleted those forgotten blog posts and stopped cringing, I considered the request. On one hand, it felt validating to have been asked. On the other, I was scared.

The deep need to live in the truth of my racial identity thrummed, but I remembered the struggles of my childhood and felt wary about summoning them once more. The show reached a lot of people and could open me up to exactly the kind of harassment I was trying to avoid. Stuck somewhere between the two points, I tentatively agreed.

I sat at our kitchen table, sweeping stray crumbs into small piles between sips of tea. A young male researcher had called to ask me preliminary questions.

'Have you lived in Scotland all your life?'

'Yes. I was born here.'

'Where were you born?'

'Broxburn, just outside Edinburgh.'

'And you just moved from Edinburgh? How long were you there?'

'Ten years-ish.'

'Were your children born there?'

'One in Edinburgh, one in Guernsey.'

'Guernsey? Where's that?'

'The Channel Islands? Off the coast of England towards France.'

'Ah. Jersey?'

'Jersey is another island down there, yeah.'

'Is that still UK, or . . .?'

'Still UK, yep.'

'OK. Great . . . Two kids?'

'Yep, two girls.'

'Lovely. How old are they?'

'Jo is thirteen and Nell is three.'

'And do they look normal?'

After a pause, voice shaking, I stumbled over my words, umming and ehmming, brain unable to rebalance. He moved on, cheerfully, to the next question.

Interactions like this always left an unpleasant taste. I felt unable to shake my disappointment in the world and was tortured by profound disappointment in myself, and my silence. This kind of thing could affect me for months, even years. I felt bitterly angry, too. A national radio station, interviewing people who had experienced racism, was perpetuating racism.

I dug my laptop out from the pile of electrical devices and wires beside my bed. Cross-legged on the crumpled duvet, I spent twenty-three minutes watching the cursor move back and forth a few spaces.

All racism, no matter how casual, immediately pulled me back into the parts of my body and mind that had never felt safe. And, from experience, I knew that the world didn't really care. People preferred to be judged on their intentions rather than their actions. I was expected to extract myself, shaking, from a tangle of racing thoughts and engage in a way that considered this young man's feelings even more than my own.

Three drafts later, I had four pages of explanations that circumvented the truth. Ten drafts, and four hours later, I'd managed to get it down to this:

Hi,

I just took some time to process our conversation and the direction the show is going in.

There were a few things you said that played on my mind, particularly you asking if my 'children look normal' (white). Now I realise for you this was likely a foot-in-mouth situation, but for me it's just so typical of what I am used to hearing. 'Innocuous' things that are said so easily. Sadly, these careless things represent a core part of the collective consciousness here in Scotland when it comes to race, and a huge part of the problem we face.

Since I only want to talk about racism on my terms at this point in my life I don't think the show is for me.

I wanted to let you know as soon as possible as I know how difficult it can be to organise these things.

My sincere apologies and good luck with the show.

Kristie

Too polite. Too afraid.

By October, we had been in Perthshire for four months. The novelty of the bucolic idyll had begun to wear off and I noticed a familiar, creeping dissatisfaction. Rural life was quiet, spacious. It felt a little baggy on me.

Instead of committing to the change, I slipped back into the tailored lines of the city, returning to Edinburgh whenever I could. But the city reminded me why I shouldn't have moved to the countryside, and then the countryside reminded me why I hated the city. I felt restless, unable to settle, caught between two lives. It was a special kind of purgatory.

In Perthshire we took the kids to see the salmon jumping at Buchanty Spout. We climbed Kinnoull Hill, the A90 twisting through the fields below us, and explored the eighteenth-century folly. 'A castle!' shouted Nell. We enjoyed Lady Mary's Walk, threw sticks into the still, dark waters of the River Earn and watched them keep our pace before disappearing into the distance. Utterly lovely, it all felt a little superficial, like small talk.

This bothered me. The land and I weren't just neighbours thrown together in conversation due to our geographical proximity. We were old friends. The kind who bump into each other after five years, say 'let's get a coffee' and genuinely mean it, look forward to it.

What I needed, obviously, was a baptism of fire.

Despite having never attempted something like it before, I felt confident that I could 'bag a Munro'. For the non-Scots, 'bagging a Munro' is a sexy way of saying 'climbing a mountain'; a favoured activity among our more motivated citizens. Perthshire is home to twenty-eight Munros, but the most local to me was Ben Chonzie. Described on Wikipedia as having 'relatively few distinguishing features' and 'often regarded as one of Scotland's least interesting', it seemed well within my reach.

Setting off, Janek at my side, I felt better than usual. I had abstained from drinking the night before, intuiting that nausea and palpitations could be a hindrance to hiking. It was a cold but bright autumn day and Scotland's tired landscapes declared hidden ardour in the sun. Brown became bronze, yellow turned to gold, and rust to shining copper. I was doing this.

Ben Chonzie is the sort of hill you first have to walk into. Its long, rolling slopes and unhurried ascent mean that for much of the climb, all you can see is the mountain itself. The path was good, but the hill looked a lot bigger than I'd imagined. Repeatedly, anxiously, I traced the track ahead with my eyes, following the twisting line until it disappeared somewhere near the summit. We walked. And walked.

An hour in, my legs were begging me to stop, and without an expansive view to reward my efforts, I was considering it. I took another 'water' break.

Janek had kept pace in a generally patient way, but now he walked ahead, stopped and, looking at my feet, watched me take every step until I caught up. I made my apologies, a 'sorry' placed between each heavy breath. What they don't tell you about bagging a Munro is that it's uphill the whole way.

After two hours, the climb changed. Steeper, less clear, less solid. People with walking poles and specialist trousers had been passing us in both directions, but now, as I crouched, in jeans, more hikers were on their way down. I mustered a 'thank you' to every 'You're nearly there, chin up!', 'So close, you can do it!' that was handed to me. Each kind word was a baton I did not want to carry.

As I continued, legs shaking, lungs stinging in the cold air, I fought a rising sense of rejection. I'd expected to be welcomed back to the land with open arms, a long-lost daughter returning home. I'd refused what the city could offer me, I'd chosen this place, and the land was still treating me as if I were a stranger. Maybe I'd been gone too long? Maybe I didn't belong here either? I felt a lump forming in my throat.

Climbing was slow and without pride. Sometimes I took only three steps before having to sit. Sometimes I pulled myself up, handfuls of heather pressing their tangled patterns into my palms. I grunted. People politely averted their eyes. There was a minimum of skin on my right heel. But if I was moving forward, I didn't care how.

The mountain didn't care either. And, as if in recognition of my determination, it rewarded me with a view back along the path we had climbed. I saw it, mid-step, and turned to gaze out from a grey boulder that looked like a very large, very perfect loaf of bread. There were clouds but the light was good. I unhooked stiff arms from tight backpack straps, lifted the camera from my bag and took my first photographs of the day. And a moment to appreciate how far I had hiked.

Ahead, Janek was walking in his familiar stride. Suddenly feeling greedy, I knew I needed to be the first one to the summit cairn. Unable to announce this absurd and pointless (but very real) desire, I instead adopted a style of Olympic speed walking I'd seen on TV as a child. Core and glutes tight, feet close to the ground, hips swinging in restrained circles, I tried to pass. Casually. Sensing me nearby, he slowed and moved a little to the right.

The moment I realised I could make it, I ran.

Quadriceps pumped and burning, hair wild, every loping step a spring into the next, I had no capacity for thought. Only cairn.

Approaching the summit, I wriggled out of my backpack and leaned it against the stacked stones' weather-beaten faces. Turning my own face to the sky, I closed my eyes and smiled into the hospitality of the sun.

I stretched, arms above my head, and walked towards the edge of the plateau until the world below was visible again. As far as I could see, peaks and slopes blanketed in a light that was taking on the amber heaviness of sunset. I had time to mouth a small, silent 'thank you' before Janek arrived.

We didn't have long. In Scotland, light fades fast late in the year. Our backs against the stones of the cairn, we drank tea quietly, a little smile on Janek's face letting me know that he knew that I knew that he had essentially waited for me for three hours and then let me win the race. I was too happy, and too exhausted, to engage. I let my mind wander.

I was surprised to realise that I wanted a photograph. Of me. Up here. But I couldn't ask Janek. He might ask questions, or tease, and I felt so sensitive, the request so silly.

'Well,' he said, pushing himself up to standing.

'I'll take a photo,' I offered. 'Of you, over there?' I gestured to where I had stood earlier. He walked, then turned to face the vista, as I had.

'Like this?' he said.

'Yeah, great.' I took the photo. 'OK, I'll jump in.' Offhand, unconcerned, I told him where to stand and showed him how to frame the photo. The shutter sounded.

'You wanna see it?' he asked, but I shrugged it off, not wanting to betray how much I cared.

I took the camera from him and put it back in its black fabric case, pressing down the Velcro and hoping the picture was what I needed it to be. Kneeling, rucksack in hand and with a deep feeling of achievement, I took one last look at Scotland's least interesting, most personally transformative view.

The light was low now, and with the air cooling quickly, my brain began to return to its chemical baseline. I had not given one thought to the fact of descent, and as we headed back, I soon realised that what they don't tell you is that it's downhill the whole way. Muscles fiery, knees protesting at the steep angle of the slope, I didn't allow my eyes to trace the path down to the car. All I needed to consider was my next step. I lifted my eyes to what was around me. Fingertips of golden light clinging to the peaks.

•

Since we had left Caithness to pursue our own lives, my brother and I had remained distant. Even during the years we had lived a few streets apart in Edinburgh, we'd hardly spoken. So I was surprised when Matthew got in touch asking if he could visit us at our new home in Perthshire.

Although Jo was in high school, she barely knew her uncle.

'Are you the oldest?' she asked, hovering behind me in the kitchen.

Squeezing past her dancing feet to the sink, I replied, 'Yeah, by eighteen months.'

'So you played together? When you were little?'

'Of course. It was just us, so we sort of had to.'

'What was he like?'

Pausing in my clearing away of potato peelings, I felt a pang of grief. 'Umm. He was sweet and silly, loved Michael Jackson, SuperTed and Teenage Mutant Ninja Turtles.' I laughed. 'He used to say he wanted to be SuperTed when he grew up.'

'Who's that?' asked Jo.

'He was a teddy that got chucked out of the factory for being defective and then was found by an alien called Spotty who brought him to life with cosmic dust and turned him into a superhero.'

Jo was confused. It took a few tries to clarify the details of what had seemed to me, up until this moment, an entirely reasonable story.

Then she was smiling. 'Mum, that's so cute. He thought he could be a teddy bear?'

'He desperately wanted to be SuperTed. He used to shove a tea towel down the back of his T-shirt, like a cape.'

'What about when you were older?'

Hands under the tap, I replied, 'When we were about ten, he was really into tractors and the Spice Girls. His favourite was Baby Spice, I think.'

Jo squealed. Baby Spice was her favourite, too.

•

After the Ben Chonzie climb, blisters dry and able to move without wincing, my bond with the land did indeed feel forged in fire. We had seen the best and the worst of one another and, reunited at last, we had a lot of catching up to do.

Perthshire is pastoral, rugged, rolling, vast, lush and barren. A little bit of everything. The house we'd bought sat at the edge of an expansive stretch of flat farmland between Glen Almond in the north and Glen Devon to the south. Faded, scrubby, earth tones in the dead of winter; bold, life-giving terracotta once ploughed; and then almost impossible green and gold at the height of the growing season. One day, shod in too-big boots, I found myself alone, tentatively exploring the land around our home.

In the city, I'd worn headphones, listening to music whenever I left the house. They had provided protection not only from sensory overload but also awkward interactions with neighbours, passersby and mums at the school gates.

As I walked along Perthshire's winter hedgerows, I felt my left hand reach up, unhooking one headphone. I listened to the sound of ice breaking under weight pressed evenly and deliberately on the frozen surface of a shallow pothole. Soon, my other ear was liberated for the crunch of crop stubble beneath my boots.

•

Lindsey had lived in Perthshire for a couple of years longer than us but had so immersed herself that she had all the airs and graces (and knowledge) of a local. I followed her as we left the main track at Sma' Glen and climbed to the plateau between the River Almond and the hills above. The ground in the longer grass remained boggy from rain a few nights before and I felt the familiar feeling of suction beneath my boots. Although I couldn't see it, I heard running water. Over the years, the hills had shed rocks onto the grassy table below and, at first, I couldn't see the structure within all that natural disorder. As we walked closer, I identified the ruins of twenty-two buildings, a corn-drying kiln and various agricultural enclosures.

Craignavar is one of a few special places where the past hasn't yet settled into the ground. Instead, it sits in the air around you and in the proud piles of fallen stone. Nothing more than a feeling, yet as tangible as the rough, moss-covered surfaces of the rocks.

I hovered beside still-standing drystone corners and gable ends. Feeling unexpectedly vulnerable, I resisted the urge to touch. As Lindsey showed me around I repeated the only phrase I trusted to convey my awe but remain generic enough to conceal my feelings. 'Wow, oh wow.'

As a lonely child so often roaming the land, I'd daydreamed of the stories stones held. Trousers tucked tight into my wellies to stop the woodlice from wandering, I would sit on the ground and place my hands on the stones. It felt as if I could see through the eyes of the person who'd put them there first. And all who had placed their hands on them since. And, just for a moment, I had found respite in other lives.

Standing among the ruins at Sma' Glen, I felt close to that again.

After Craignavar, I began to notice drystone all around me. In Perthshire, there are hundreds of miles of walls, and it dawned on me that, despite my fascination, I had no idea how they were built. Or by whom. That's the thing about Scotland's drystone walls: they're there and feel like they always have been. As if they were created, not by diligent human hands, but by the land itself.

My curiosity led me to blogs written by those who were similarly intrigued, a few old newspaper articles about wallers working on this thing or that, and the discovery that a drystone 'beginners' weekend' was being run a mere ten minutes from our home. And I could register for the course right there, online.

I didn't even have to call anyone.

Smiling to myself, I decided this must be fate, a universe-approved path forward. The kids were playing quietly and only looked up as I walked by to get my purse from the hallway. I sat back down and entered my credit card number. It cost more than my monthly wine budget. But, I thought, closing my computer, you can't put a price on destiny.

•

My brother sitting at our kitchen table was a lovely thing. He'd brought gifts: a Thor action figure, a set of wide-eyed pastel ponies and good Chardonnay. We drank and chatted. As we worked our way through the first bottle, I saw his body loosen a little. When he casually referred to me as 'Krist', a nickname that only he ever used, I smiled to myself behind my glass. Halfway through our second bottle, the music was on, each of us commanding Alexa to play the greatest hits of our childhood. Whitney Houston, Go West, Cat Stevens. We laughed hysterically, remembering how we used to say our mum's car was possessed by Peter Gabriel because 'Sledgehammer' would blare from its speakers every time she started the engine.

Matthew lifted Nell onto his knee. I watched, my face stoic, as he enjoyed her attempts to pronounce 'crackers'. Busying myself with dinner prep, careful not to draw attention, I listened to him enthusiastically discuss Marvel movies with Jo. My children. My little brother. Hearing them pester him – 'Uncle Matthew' this, 'Uncle Matthew' that – my heart was as full as my glass.

In the coming months, we spent several warm evenings sitting in the garden until late. Talking about childhood, and life, we got to know one another. I was surprised at how open and kind he could be. If I even hinted at being tired, he would protest loudly (while I shushed him for the neighbours' sake), then pour me another. We went to bed late and drunk, never quite managing a hug, and he would be up and away before any of us had woken. This was our pattern, and although our communications between visits remained sparse, I saw my brother more after we moved to Perthshire than I had in all the previous twenty years.

•

I counted down the days, announced it on social media, talked about it constantly at home, and then, the morning of the drystone course, I was 'ill'. I said I had a headache, the beginning of a virus. I said I would take the course later. I said it wasn't responsible to make anyone else sick. But I knew, and they all knew, that I had a hangover.

Sitting on the edge of my bed, Janek reached out to feel my forehead then thought better of it. He encouraged me gently. 'You should go. You'll feel better later in the day, and if you don't, I can come get you.'

Annoyed, I dug in my heels and fully committed to having the flu.

He paused, looked off into the distance, and, sensing what he was working up to, I rolled my eyes, waiting for the inevitable.

'Why don't you ask Luke if he wants to go instead?'

Throwing back the duvet so he could see my whole face, trying to ignore the thumping at my temples, I sniped defensively. 'You can tell me straight-up that you think I'm wasting the money, Janek.'

I didn't want anyone else to go on the course, I wanted this to be my thing, but I also wanted to stay in bed. Eventually, the guilt found its way in and, from the overheated twilight beneath my duvet, I called Luke and offered him the place. He accepted. Underneath the nausea, I felt the stirrings of disappointment and jealousy but instead of dealing with any of it, I stayed in bed all weekend doing my best impression of someone who was dying. By Sunday evening, I'd even managed to convince myself.

•

Once again hungover, I was on the train to meet my friend Amy in Edinburgh. A book I'd seen recommended on Twitter years before popped into my head. I found an audio version of *This Naked Mind* by Annie Grace and downloaded it to my phone. The first chapter was compelling, about how we're conditioned to see alcohol as innocuous, even beneficial (think antioxidants in red wine), but really, it's a net negative.

Since moving to Perthshire, alcohol had been affecting me more. My period had become heavier, staining my underpants and thighs a bright Sharpie red several times a day. I'd found myself breathless from long sequences of heart palpitations, was three stone heavier and, despite being outside a little more often, I was more unfit than I had ever been. I'd felt anxious and hungover even on days when I hadn't partaken the night before. And my parenting was ruled by my mood, which is to say, ill-tempered.

A middle-aged man with a messenger bag slung over his shoulder walked down the aisle towards me, scanning the carriage for eye contact. Territorial, I put my hand on the seat next to me and set a misanthropic look on my face. He kept walking. I returned to staring out the window.

After years of going to sleep drunk, my drinking was largely habitual. I cared little about what I imbibed, though I maintained the illusion that I was a budding sommelier. The habit was expensive, so if we didn't have a guest, I drank whatever was the next step up from Buckfast. I didn't derive much pleasure from it, I simply couldn't remember what my life was like without alcohol. Worse, probably?

As I exited the train I made a mental note to listen to chapter two on the way home.

The unassuming wine bar was set among the off-licences, cafés and betting shops at the top of Queensferry Street. We sat in the window, framed by dark wood, our table decorated with a single candle and a tiny bowl of olives. As we chatted, catching up, edginess and nausea forced me to angle my chair away from bright streetlights and towards the softly lit interior. I ordered an organic Viognier from France, and at £12 a glass, I hoped its good quality would be enough to quell my queasiness. I drank the wine quickly and resolutely. A liquid weight to force down vomit.

The nausea dissipated slightly after my second glass, and although I felt relieved, I suspected I'd pay for it in the morning. A third and fourth glass followed for no reason other than to get as drunk as possible. Walking to the train, the air felt warmer; Edinburgh's lights had never looked more beautiful and were twinkling just for me. On the journey home, with thoughts of sobriety gone from my mind, I listened to music instead of the audiobook and indulged in my favourite drunken activity: imagining a perfect day.

A day in a life where my husband understands me and the kids walk into the kitchen and exclaim, 'Eww, Mum and Dad, stop kissing,' and we all laugh. A life where I am methodically working my way through a detailed to-do list, strength training, able to tolerate the sensation of wearing a bra, and just about to finish my second PhD.

The next morning's hangover was overshadowed by tedium. So weary, so stuck, I caught the beginnings of a resolution forming. Distracting myself, I pulled together the mental scraps of my chaotic way-too-late-must-do list. I hoped that the stress it caused would be enough to divert my attention. I had to be careful. The thought was fragile; if I tried to shape it into anything, it would break.

Over the course of the day, that feeling turned into something almost like a purpose, and that almost purpose turned into some semblance of a plan. I never finished the audiobook, there was no big announcement, no declaration of my intentions, especially not to myself. The less I knew, the better.

The first week was fine. My brain graciously accepted the short break in proceedings. Week two, my body began to protest. At first quietly, a general vibration under my skin. I lost my appetite and woke each morning sweating like I had the flu. Then loudly: nausea, dehydration, dizziness. But the worst was yet to come.

For more than ten years, alcohol had been my principal method of dealing with stress and sadness. It had provided relief, which I had mistaken for catharsis, and in turn mistaken for closure. I really had no idea how much I was still carrying.

It was a late-night onslaught, a crossfire of insecurities and painful memories. The shame in the realisation that I was an alcoholic. I resisted, reinforced with the robust defence of sleeping tablets, and spent my mornings post-sedative groggy. A new kind of hangover, comfortably familiar. I fought the war the next night, and the next, and the next. It seeped into my mornings, afternoons and, before long, every waking minute.

I'm not sure I believe that any of us have a soul, but if anything could convince me, it's the writhing anguish of a grieving body. Truly, nothing could have prepared me for the grief that came flooding in. Disoriented and barely able to function as waves of regret and loss crashed over me, I struggled to know which way was up. I had lost friends, my father, my grandmother, and I had never mourned their deaths. But the grief I felt was expansive. I mourned for everything.

My broken family and what I'd been through as a child. I thought of uni and Daniel, the abortion, the miscarriage, the self-harm, the rape, and every other time I had been mistreated by a man. I wept for failure after failure, my own, my family's and the world's. How I'd never had a fucking chance because my parents, and their parents, never had a fucking chance. How I'd tried to do better for my children but I wasn't sure I'd managed. How I'd greeted them with closed doors and resentment when they needed me. How I'd hated myself and still hated myself. Wide open to the churning past, the horizon hidden by swells of self-loathing, my thoughts were more than the usual intrusive ideation. I truly wanted to die.

Shut in my room, pacing, sobbing, sleeping, unable to handle even the light coming in from outside. One drink, right now, could make it all go away. I bargained with myself. *Who goes cold turkey like this?* I could do it incrementally, ease into it. In fact, wasn't I more likely to succeed if the process was more regulated?

I could sense the bottles in the kitchen. Tequila, in particular. A 40% ABV enchantress. I walked over and stood at my bedroom door. I looked at the handle and willed it to move on its own so it wouldn't be me fucking up and going downstairs and giving in. Through the door I heard Nell singing her version of the Moana song while she rummaged in the snack drawer.

Hey, it's okay, it's cake, you're welcome.

Staggering back to my bed, I was overtaken by the largest silent sobs. I had two very good flesh-and-blood reasons to recognise that any internal arbitration was the addiction talking. My phone was under my pillow. Texting Janek, I told him to go to the kitchen and throw out all the alcohol. Now.

I had to keep going precisely because it was so hard. I never wanted to do this again.

•

Getting through a day without booze was excruciating, impossible, and so I started measuring my time in smaller and smaller increments. *Make it an hour, half an hour, five minutes – make it one measly fucking minute.* I tried everything I said I would never do. Affirmations, meditation, yoga, but nothing would stick. I was simply working through all the necessary evils, looking for deliverance.

The first run of my adult life was disgusting. The pavement was trying to punch its way through my legs to my heart, which, terrified, was beating faster than it ever had. Breathing was painful, and my body was turning itself inside out, innards twisting and pulsing their way towards my throat. Looking at the timer on my phone, I couldn't believe that thirty seconds of anything could have this effect on a human body.

As awful as it was, while I was retching in a bush or bent over struggling to catch my breath, I wasn't telling myself I was a piece of shit. Nor was I thinking about getting drunk. It was good for me.

It was good for me, so I restricted it. To avoid injury, burn-out or even boredom, I proceeded cautiously and with purpose. At first I ran for those thirty seconds. Once that felt tolerable, I managed one minute.

Every run in that first month felt awful, but I was still sober. Soon I was running for two, three, four minutes at a time. I bought a full running kit: a special, lightweight, glow-in-the-dark jacket, a harness to carry water and my phone, three pairs of trousers, two pairs of trainers. Absolute necessities for my half-mile runs around the village.

•

Jo wanted to consult Matthew about her list of 'Ten All Time Best Superheroes'. So when he texted asking if he could visit that weekend I replied, 'Yes! But don't bring wine, I'm aff the booze!' and waited for one of his usually swift responses. It didn't arrive. On the Friday, I messaged to ask if we were still on for the weekend. He replied before my phone screen had a chance to darken, saying that something had come up, he wasn't going to make it. I was disappointed but I understood.

A few weeks later, he made the same enquiry, and I told him that he was welcome any time but that I was two months sober, so to hold off on bringing alcohol. He fortified his response with his long-held ability to persuade me to do anything.

Oh, come on, Krist, how often do we see each other? I'm sure you can make an exception for one night.

The whole point of sobriety is that I don't ever make an exception.

The whole point of sobriety is that it makes you a loser, sis.

I remembered how desperate I'd felt as a child for him to like me. Paying him to hang out with me. Climbing the gate that said 'No Entry', or opening the hidden Christmas gifts we'd found mid-September in the hall cupboard. I stood up and physically shook my body, trying to remove the clinging droplets of his words from my mind. It was a choice. Sobriety or a relationship with my little brother. Or was I being dramatic? I joked my way to a boundary.

Well, I've always been a loser, you know that. Would love to see you but I won't be drinking . . .

My brother wasn't the only one who struggled to accept my sobriety. At least three friends were unwilling to let me use the word 'alcoholic', debating me on the definition. Others resolutely neglected to remember that I was sober: bringing me wine as a hostess gift, insisting I come to their birthday party in a wine cellar. Amy, who had been there when I'd had my last drink, asked me several times to join her in getting shit-faced because she was so stressed with work. When I declined, they all said the same thing: now that I'd proven I could do it, why continue?

My goal wasn't to convert the masses; sobriety was private and precious. I gave my reasons but applied no pressure, passed no judgement. Still, articulating what alcoholism looked like for me was enough to elicit curt tones and a defensive stance from them. I'd try to laugh it off, sharing an anecdote from my growing repertoire about rural life. All mud and no buses, amirite?

But soon, the fears I'd had about leaving Edinburgh became realities. Invites slipped away, then more friendships. Life as a sober, mixed-race woman was, I discovered, exceedingly quiet. The thing was, now I didn't really mind.

Sobriety was the gift that kept on giving, especially during that difficult first year. After the emotional agony, I enjoyed daily reminders of why it was absolutely the right choice. My period was more manageable, I slept well, my palpitations stopped, my anxiety diminished, my depression lessened, negative thoughts too. I lost loads of weight. There was one thing, however, above and beyond all else, that kept me going. I was a much better parent.

After Nana's death, my mum told me how she had learned, as a child, to listen for the tell-tale signs of her mother's hidden drinking: a slight slurring or sharp tutting sound she would make with her tongue. My mum had battled alcohol too, and I remembered (but didn't tell her) the fear of not knowing which Allison I was going to meet. All versions felt oppressive, even the happy ones. The drunk mum who wanted you to sing songs, dance around the kitchen or have big conversations still didn't really give you a choice.

Then there were the scary, explosive versions. Now that I was thinking more clearly, I could finally admit to myself that I had lived in constant fear, trying to ascertain how, and who, to be. Watching with worried eyes, asking, 'Are you OK, Mum?' 'Did I do something?'

It had been my mother as a child, then me as a child, but now it was Jo and Nell, children, inheriting hypervigilance. Asking the same questions, getting the same dismissive answers.

Every day that I maintained my sobriety, I was extracting my kids a little further from that toxic guessing game.

•

Since I'd reclaimed custody of Jo, Luke and I had moved in opposite directions. I was guided by a deep feeling of responsibility to my children, and he was running from the very same.

I had always been honest with him about how his behaviour was affecting his daughter, but I'd lied to Jo. 'He's working a lot, honey,' I'd say in my most regretful tone, knowing he was in Greece with his latest twenty-two-year-old. 'Dad's not feeling well,' I'd tell her, barely able to contain my anger as hurt filled her eyes. I gave him a hard time. He said I was a bitch. I sent him links to books that would help him get his shit together; he told his friends I was mental (true) and overbearing (true), without telling them that I was right (the most true). I recommended therapists, and he told me I needed one more than he did, which was great because I already had one, would he like her number?

There were times when we could co-exist, managing our co-parenting responsibilities through polite texts and handovers. There were times we would shout. Or the time when, six weeks before I married Janek, Luke told me he was taking me to court for full custody of eight-year-old Jo.

I was livid. I pointed out that he couldn't seem to spend regular time with his daughter even on weekends, he lived with four men in a shithole apartment, and slept all day after his night shifts. I told him he was an arsehole. I told him I would hire a lawyer. I told him I had moved on and he should be grateful another man was willing to do the job he wouldn't. For some reason, this had not endeared me to him.

After months on the custody warpath, he'd decided that he had scared me enough, or maybe he'd grown bored with the idea, because he just never mentioned it again. Life continued on as it had before. I wouldn't say that I'd come to accept his inconsistency, but I had resigned myself to it. Our shared dark sense of humour didn't hurt and, after years of animosity, our relationship had slowly become patter-based. Sure, superficial, but if communication had to happen, this would do. Getting along was better for Jo.

Annoyingly, Luke had enjoyed the drystone course. So much so that he continued to learn the craft. I watched him, doing a poor job of hiding my envy. I teased him, saying that he'd appropriated my specific and fervent lifelong dream of becoming a drystone waller. I was only half kidding.

Out in all weather, he dismantled and rebuilt, slowly graduating to simple wall repairs, eventually sitting his level one exam, and soon after his level two. (Yes, you can sit drystone walling exams.) Now, working full-time as a waller, he seemed at ease. More connected with nature, more connected with Jo, more connected with himself, maybe. Fucker.

Cup of tea in one hand, index finger of the other flicking across his phone screen, Luke explained what he was working on.

'It's a retaining wall, thirty metres long, shorter than standard height. Not sure about the copes yet, maybe double copes.'

Squeezing honey into my own tea, I smiled. Copes. I had no idea what he was talking about.

Silence.

Distracted by the need to rescue a structurally unsound biscuit, he snorted without looking up.

'Wanna see?'

'The wall?' I asked, suddenly unsure of my place in the conversation.

'Yeah.' His phone was back in hand, finger tapping the screen, crescent-shaped shadows of dirt beneath his nails.

'This is the one I'm working on. The wall I was telling you about. Retaining, low, blah blah blah.'

Before I had a chance to respond, he flicked to the next image.

'This is like a wee pillar thing, you can't see it but . . . hang on . . .' Turning the phone to face him, he swiped several times before turning it back to me. 'This is the sun dial that's on top.'

It was so nice to see him like this. Watching his face as he showed me more photos, I could see how proud he was of his work. And so he should be. Literal piles of stone transformed into beautiful structures. I wanted to do that. I badly wanted to do that.

'Amazing, Luke,' I said, making sure to lower my head and catch his eye so he could see that I wasn't being sarcastic.

He shrugged, 'I mean, you could do it if you wanted . . .'

Before he had even finished, that anxious British politeness kicked in and I accidentally interrupted with, 'Oh, I wouldn't want to get in the way, it's your job and it would be a lot to teach . . .'

I waited for him to speak.

'Well, how about the garden? That space you're not sure what to do with. We could dump some stone in there. Have a play.'

•

Watching Kirsty try to load medium format film in the rain, I was worried. For the first time, my life was quiet and good. I was sober. Sitting on the grass at Sma' Glen, looking out over the river below, I hid anxiety behind that classic pose of strength and serenity: the smize.

I heard Kirsty's voice. 'Can you look straight ahead?' And out of the corner of my eye, I saw her yellow coat, a vivid smudge of colour against the muted green and brown of the landscape. Doing a good job of looking straight ahead, I was glad I had chosen this spot for the portrait. But the presence of the hills, heather and nearby Craignavar couldn't completely soothe my unease.

I'd followed Kirsty Mackay on Instagram for years and it felt like a big deal when she followed me back. One of my favourite photographers, she had asked if I would be the subject of her contribution to a book about concepts of Britishness in the post-Brexit landscape. Immensely flattered, I replied to her with a message containing too many exclamation marks, agreeing to a portrait! and an interview!!!

I was light-skinned and mixed-race, and once the book was out, the whole world would see me. I was afraid of being harassed for being mixed-race, but I was also afraid of being dismissed, again, for not looking the part. For not being 'brown enough'.

But I'd always been brown enough to be called a 'Paki'.

Suddenly, I had a keen sense of my body. I shifted position, adjusting my posture. Still looking ahead, I panicked. *Was I slouching in the other pictures?* But Kirsty was finished, telling me I could move now and packing up her bag.

•

Each year, ploughed fields in Perthshire give up a new bounty of boulders. Surplus to farmers' requirements, these stones are either dumped down the side of whatever field they came from or, if they are particularly numerous, collected in one designated area. With wallers in the mix, one farmer's rubbish becomes another man's treasure.

The ability to attune my eyes to the stone piles happened quickly, and I found myself engrossed in a task that sounds like a game your mum makes up and then enforces on a road trip.

Once 'we' (I, yelling) spotted a pile lying in a field, we would go to the nearest farm and check if the stones belonged to them. Often they did, but sometimes we would be pointed over a hill to another farm, and off we would go again.

The whole process involved a lot of farm tracks and a lot of following directions given by people who knew the roads so well that they didn't feel the need to be specific. Then, once we had a farmer's permission and had agreed on a price, we filled Luke's van and took the stone home.

I was going to be a drystone prodigy. My childhood connection to the craft, and to stone, had primed me with a special intuition that would guide my hands in building things no one had ever seen before. But by the end of my first afternoon, I had laid six foundation stones and walked away from the wall in frustration at least fifteen times. I had begun to shout at Luke about his teaching style, complaining that he was 'making it about him', but then demanding he pay me more attention when he backed off. The stone wasted no time in humbling me.

Early the next morning I woke in barely-there light. Although Luke wasn't due to arrive for another hour, I got up, got dressed and went outside, careful not to let the kitchen door slam behind me. I'd hoped to feel better about the work I'd done, but the six stones I'd placed the day before sat there marinating in mud and my incompetence. Looking back at me, their dawn-time grimaces were disapproving and expectant.

Deciding that the foundation must be an especially tricky part of the structure, I picked up a stone near my feet. Then, remembering what Luke had told me, I placed it over two stones, with its length going into the wall, and watched as it tipped to one side, now barely covering the join beneath. I moved it around, not-so-gently offering it a variety of positions to settle in, and that's how Luke found me twenty minutes later.

Tyres on gravel had signalled his arrival, and I felt him watching me as he changed into boots and pulled on gloves. He had been staring for ten minutes when he finally extended his hand. I passed him the stone, dead-eyed. It wasn't the right stone for that part of the wall. What could he try that I had not already?

'I've already tried that,' I said, my tone tight with annoyance.

'Hmm,' he said, focused, barely listening.

I crossed my arms, stood back and looked smug, waiting for him to fail. His eyes were on the wall, then on the stone, then on the wall again. Reaching down, Luke gently dropped the stone. It made a heavy 'click' as it settled into place, and he whipped back towards me so fast that I didn't have time to hide my incredulity.

'I just turned it over,' he said.

•

In the months after Kirsty took my portrait my chin became a bed of angry red patches, the outer third of my eyebrows lost to busy fingers. I slept restlessly, dreaming. Whether monumentally boring day-to-day scenarios, Pedro Pascal once again rejecting me, or being chased through underground tunnels by Nazis, there was always, somehow, a glass of wine in my hands. I would wake up, heart racing, filled with horror, convinced that I was no longer sober.

When the book was released, I shut my ears to what the world had to say. I shared the photos on my private social media accounts and didn't read any public comment sections – I didn't have the energy to debate my existence. But I agreed to go on a podcast. During the recording, I sat on my bed and answered the mostly straightforward, biographical questions. As the conversation wound down, the host asked if there were 'some spaces of hope' to be found.

I listened to another guest answering so eloquently, so diplomatically, and I felt something building inside of me. It was wilful, organised and expanding rapidly. My turn came, and I was talking before I had given my brain permission. Laughing nervously, prefacing with caveat after caveat, I could hear my voice shaking, feel my tongue moving slowly as I spoke.

'So I think the way that hope is sometimes presented to us, and it can be presented sort of through a political lens too . . . I'm a bit wary of it sometimes because, I think hope, even at the best of times, is sort of an intangible thing and I think that, you know, in terms of making real change, I think what we need is, like, something more solid, pragmatic, all of these things, and something very tangible . . .'

How many times had I said 'I think' and 'ehm' and, oh my God, was I shouting?

'. . . and also, I think that hope can be used as a sort of stall tactic for change, where we leave people feeling "hopeful", waiting, and I . . . you know what, I think there's been enough waiting.'

Not quite as polite. A little less afraid.

•

The mechanics of a Rubik's cube can be easily explained, but the skill of working with one, understanding and then solving the puzzle, can take years to learn. Although each cube is solved using the same mechanisms, following the same rules, each problem is distinct. Drystone basics are easy to grasp, but the ability to, for example, quickly see the stone that works next to another, or create visual balance and flow, is developed through experience. I learned in those first months of experimentation that drystone walls are multi-ton 3D puzzles, solvable in a billion different ways, but not with your mind or hands alone.

My brain had always struggled to get on board with tasks that other people said were normal or necessary, and that I described as 'mundane' or 'meaningless'. Secretly, I had been afraid that nothing would be more mundane than stacking stones. But as long as the puzzle infuriated me, I was captivated.

There was a meditative component, too. A state of deep focus, so I didn't notice how tired I was until the end of the day. It made me crave water, not wine. And I found that I liked going to bed exhausted, too tired to think of anything but sleep, my body grateful for rest.

I'd felt perpetually worn-out since Nell had been born, but

now I had more energy for the things I wanted, and needed, to do. Between sobriety, running and drystone, my physical and mental health was the best it had been in years. Sobriety had been especially instrumental. But it changed my marriage.

Drunk, I had been numb to the many things that bothered me sober. Without the anaesthetic, Janek and I argued incessantly. I denounced his prison camp vibes parenting style; he said mine would raise lazy, unimpressive daughters. He said he shouldn't have to do more around the house after working all day at his very important job; I said I worked all day looking after the kids, being a photographer, learning drystone and occasionally making the executive decision not to kill myself. He called me an idealist; I called him a Tory. We even fell out about which cupboard the cups should be kept in (the one above the kettle). Every time we fought, I felt shaky and irritable for days afterwards.

I'd assumed it was him not understanding me, a neurotypical person, and me not understanding him, a neurodivergent person. I'd read up on how to communicate with someone with ASD, but no matter what, we always ended up at odds. I'd tried everything. I'd tried so hard. I was sure the problem wasn't me.

Messaging online with the friend who had put me onto the antidepressants that had saved my life years earlier, I complained that it was tiring being neurotypical and trying to understand neurodiversity. She wrote back, sympathising. Then another message. Just one line, followed by three laughing/crying emojis: 'Except you're def not neurotypical.'

Huh.

Placing the Hearting

I grab a handful of hearting from the bucket next to me and push the fragments into the gaps between the bigger stones in the wall. I make sure that they fill the spaces and sit tightly. Some need hammering, but just a few taps.

On the inside of the wall, these small stones quietly bind its faces together. A hardworking core to stabilise the structure as it settles and faces the elements. In time, thoughtlessly placed hearting makes itself known in gaps, loose stones or the invasion of debris. Hearting is the unseen, integral part of drystone, and every piece must be placed by hand, with intent. But drystone walls have an insatiable hunger for small stones, and I am soon grabbing the last one in my wheelbarrow. I need to make more.

I still haven't quite got used to the way large hammers engage the whole body. So when I pick up the sledge, I shake out my shoulders and arms. Once the hammer is in the air, I feel myself tense, anticipating impact. Metal connects with rock, the sound louder through my bones than through my ears. Some stones in the bucket break, but most remain intact. I swing again, and again, and keep swinging until there's only rubble. It's exhausting, but there's no point in avoiding the inevitable. Hearting is tedious but important. A hidden resilience deep inside the wall.

Weaving between kids, parents and staff, Luke and I made our way towards the laminated 'Night of Champions' sign on the double doors. In the auditorium, I employed the same tactic I use for buses, heading to the back away from the crowds. Luke followed while Jo flitted around, saying hello to new friends, engrossed in the novelty of school outside school hours. As the lights lowered, she came bounding up the stairs to sit with us, cheeks red, eyes bright.

The back of the hall was away from the other parents and out of the earshot of teachers, so, asserting their autonomy, most of the older kids ended up sitting there too. As the ceremony began, the auditorium fell quiet. All except a group of boys sitting four rows behind us.

'Ugh,' said Jo, leaning towards me. 'The football boys.'

The first award winner was Gemma. She accepted a trophy for debating. The boys pronounced her 'a fat bitch'. The next was Dylan. They called him 'a queer'. As I turned around in my seat, Jo put her hand on my arm and, through barely moving lips, pleaded with me not to say anything. Embarrassed that I'd had to be told, I settled back, facing forward, and patted Jo's knee. I would behave myself.

Jo's name was called. She unfolded herself from the seat. My tall, elegant daughter had been named 'most improved netball player' and, even more impressively, 'players' player'. I gazed up at her, adoring her, as a voice behind us said, 'Paki slag'.

The boy said it loudly. He said it clearly. We froze and listened to the rest of the group laugh. I saw Jo hesitate before realising that, with all eyes on her, she had no choice but to continue towards the stage.

She took a few steps, then turned back to me and gave me a look that told me how upset she was but also: *Don't you dare*. I watched her compose herself before she walked onstage, smiling.

While she stood at the podium, squinting into the lights, the comments continued – 'ugly', 'smelly Paki bitch' – with more pubescent male voices joining in. Every cell in my body was ready to get up and pull those kids out of their seats. I turned to look directly at them.

They remained reclined, staring back at me, until the one in the middle lazily raised a hand to offer the universally understood sign of 'wanker'. Closed fist moving up and down in the air, his friends' bodies exploding, falling over themselves and each other to shove him in approval.

Luke pulled my shoulder to face the stage. 'She's coming back,' he hissed. 'Leave it.'

Shaking off his hand, I looked towards the stairs and had just enough time to smile broadly as Jo sat back down and handed me her award.

Putting on my best 'everything's fine' voice, I said, 'Well done! Are you going to put this in your room?'

Jo was silent and slumped lower and lower in her seat as she listened to the boys making comments about the other kids on stage. I gripped the cheap metal of the trophy in my hands and looked straight ahead. By Jo's slouch alone I could tell that this wasn't the first time.

My thoughts were interrupted by her leaning towards me, jacket already in hand. 'Mum,' she whispered, 'can we go?'

I nudged Luke, and we left as quietly as we could, ducking when we got near the stage for no real reason except to signal how much we regretted the interruption. As we exited the hall and walked down the corridor, music began to play and there was the muffled roar of hundreds of people moving together, gathering their belongings. Calling to Luke, who was walking in front of me, I gestured back towards the auditorium with my head and said, 'Back in a minute.'

I turned and began to move through the crowds suddenly all around me. I saw the headteacher at the far end of the stage and approached him with purpose. Within a few feet, I became aware of other, seemingly aimless parents around me. I lingered near the edge of the group. Fidgeting, checking my phone, waiting my turn, I watched and wondered if you can judge a person's character based on their footwear.

Some shoes say so much. His were long, pointed, patent, with a little buckle, and my brain referenced an image taken from the memory of a childhood storybook. Rumpelstiltskin dancing next to a spinning wheel.

My phone vibrated with a text from Luke. A single question mark. I put it back in my pocket.

Mr Duncan, buzzing from a successful evening, shook my hand firmly, and I really did believe him when he said, 'Pleased to meet you.'

'Hi, yes. During the ceremony, a group of boys' – I looked around the auditorium, searching through the crowd – '*those* boys.' I indicated with my eyebrows towards the double doors, making sure he acknowledged the group before continuing. 'They spent the entire evening harassing the kids who won awards, including my daughter, Joanna. They called her a "Paki" and a "slag" several times.'

I waited, gauging his reaction. No raised eyebrows from him.

We began to talk at the same time, so I stopped.

He paused, breathing in sharply before addressing me in his apology voice. 'I will speak to the boys and make sure that it does not happen again.'

•

In the C&A changing room, I sucked my stomach away from the waistband of the skirt and raised my shoulders to stop the seams of the jacket from touching my skin. My mum asked again if I was sure these were the clothes that I wanted.

Already struggling to fit in, I'd decided to look as nice as possible at the primary two Christmas party. Nagging for weeks, ignoring the secretive way Mum looked into her purse when we were shopping, I had eventually persuaded her to buy me something new. I was set on this black velvet embroidered skirt and jacket.

It looked like something that Denise would wear. Jealousy wriggled in my tummy when I thought of her blonde hair, blue eyes, petite frame. The Hawaiian grass skirts we played with in class went all the way around her waist and covered her bum, swaying perfectly as she imitated a Hula dance. And her legs were slim and bruise-free like my Cindy doll's. Everyone told Denise how pretty she was, even the teachers.

The morning of the party, I woke up and looked at my new favourite outfit, still in its packaging on the rattan chair at the end of my bed. Sliding my hand under the plastic, I touched the velvet and enjoyed its fine bristling under my fingertips. I was proud to own something so pretty. Today would be different.

I pulled on the skirt, trying to ignore the way its elastic rubbed my skin, and as I worked my left arm into the sleeve, a small shiver ran through my body when the seam touched my skin. I had to pull my arm back out and try again. Glancing at my mum, I saw the concern in her face.

'It's fine,' I protested loudly. 'I already told you!'

My mum picked the hairbrush up off the bed and placed it on her lap before straightening the shoulders of my jacket. I squirmed, snatched the brush and ran it through my hair, feeling sparks of static. Then I put the brush down and firmly flattened the remaining frizz with my hands.

'Ready?' my mum asked, moving towards the mirror. I beamed and nodded. I was indeed ready to be astounded by my sophisticated and objective prettiness.

Instead, dismay. Nothing looked how it was supposed to. My hair, still frizzy, stuck out on one side but not the other. My face was round, my nose too big, too pointy, and the dark lines under my lashes made me look like a panda. I felt my eyes fill with hot tears as my gaze travelled to my legs. Sturdy and bruised like the boys in my class. Mum caught my eye in the mirror. 'Och, you look lovely, Kristie. Beautiful!'

I shouted, 'Don't say that! You've ruined it!' First the skirt, crumpled under my stomping feet, then the jacket thrown down next to it. Standing in my vest and pants, I looked at my reflection, blurry through tears.

The Christmas party would come and I would stand awkwardly in faded black leggings and the T-shirt I always wore, watching my classmates tie themselves to each other with tinsel. My mum would try to return the outfit, but of course she couldn't. And Ms Wright, though kind to a fault, would never look at me in a hula skirt, smile, and call me pretty.

As I grew up, I came to resent all comments on my appearance. Even compliments. I didn't want to love how I looked, I wanted to forget it. I'd imagine myself as a floating consciousness. No physical form, just the essence of who I was suspended in the air. Like I was at the beach, arms out, leaning, body held by a strong wind.

Even the body positivity movement, which gathered momentum in my late teens, did little for me. It still forced me to consider the standards of others: still white, still prioritising male power and pleasure. What I wanted was stark body neutrality. To accept bodies as an inconvenient, inevitable part of being human.

But, of course, as I got older, things got more complicated. The mental space in which I gave up my physical form in favour of a levitating soul became lonely. Soon, the discomfort of being noticed faced off against the fear of not being noticed at all, and I began to push myself into the prettiest shapes. From my late teens – with the brief exception of using it to care for others in Guernsey – I knew my body solely through the eyes and approval of men.

I exercised only to look better. Weighted side bends to accentuate my waist, squats for a juicy butt, and flexibility training to do the splits, because men really did love that. During sex, I would close my eyes, feeling sorry for my partner who had no choice but to look at, and fuck, something so disgusting.

After giving birth at twenty-one, still young and strong, I believed that my post-pregnancy body was worthless. I'd love to say that men did anything to dispel that idea. One dumped me when he discovered soft pink channels of stretch marks cutting across my belly. A 'sorry but it's not for me' kind of thing. Others asked if my vaginal canal had 'held up', if my breasts were still fit for purpose after breastfeeding. If I'd considered plastic surgery.

I died inside. And got really good at giving blow jobs.

Taking up running in my mid-thirties was my first movement towards a different relationship with my body.

Drystone took me one step further.

•

After spending three months building and then dismantling a wall in our garden, I could almost put it back together from memory. Although the learning curve had been steep, certain things had become easier.

The basic rules had settled in my mind: length in, one over two, two over one, as much contact as possible, and hearting placed with intent. I had begun to understand what made a stone good or bad, in general, or in a specific spot.

Stones have characters, and I played favourites. Drawn to weird shapes, stones that no one else would pick up first, or at all, I kept them beside me, determined to find them a place in my wall. I made up names for stones of unknown (to me) geological origin. Often covered in mud, the weight of a stone was my first clue as to its type, and I knew if something was round and weighed a lot more than expected, it was likely a Heavy Cavill. Under the dirt, these stones were a vibrant and surprising powder-blue, the same colour as an obnoxious pair of jeans worn by extra-large human and actor Henry Cavill in a photo I'd seen online.

Outside building hours, I chatted constantly to Luke about walling. I wrote enthusiastic messages detailing encounters with this stone or that, sent him images of problems I'd solved and bits of wall I was proud of. Taking a new tack with him: instead of demanding what I wanted, I gave him the chance to pull his head out of his arse long enough to read between the lines. And he did. Seeing my progress, Luke suggested that I join him on a job.

A week later, morning mist not yet lifted, we trudged across a churned-up field edge. Branches and twigs littered the ground around us, casualties of the recent storm.

Today we were 'gapping', fixing fallen or damaged sections, and I was pleased to see the wall was entirely sandstone. Often driven to frustration by the rounded and irregular shapes of fieldstone, I loved to find a piece of sandstone in my pile. Flatter, friendlier and easier to place and pin, sandstone was a treat. As he pulled on his gloves, Luke offered me a small bit of wall that had collapsed under the weight of a large branch. I watched him walk a hundred metres away to a larger, more complicated repair. I was on my own.

Putting on the gloves he'd given me, I was annoyed to find they were still a little damp inside. I lifted one to my face and smelled sweaty feet. 'Oi!' I shouted to Luke as I held up the glove and mimed vomiting. He shrugged. I worked my hands into the cold, pungent material.

On the way here, Luke had explained the process to me and, remembering those instructions now, I repeated them back to myself.

Strip out, organise, rebuild.

Facing the wall, I pulled what wasn't already on the ground into the mud. Then I began to pick up stones, holding my right hand against their edges to ascertain how they should be organised. Like drinks, stones are measured in finger widths. Soon I had four piles: two, three and four fingers tall, plus a pile for hearting. A tap on my shoulder alerted me to Luke's presence.

'Tea?' he asked, and we walked back down the field to the van. Once inside and mercifully gloveless, I saw that the dirt had somehow worked its way through the fabric onto my hands. With no phone signal, I passed the time silently staring at the clock on the mud-splattered dashboard while I ate a biscuit. An hour had passed quickly. Following the direction of my eyes, Luke said, 'It's an hour behind.'

Gulping greedily from a plastic cup that was handed to me, I considered how fast time flies when you're stripping out.

Then, a bottle of water shoved into one coat pocket, three biscuits in the other, I followed him back to the wall, ready to begin the rebuild.

The foundations and lowest courses of the wall were still intact, and as I worked, I was grateful not to be lifting anything larger than a phone book. The sandstone, already shaped, went back into the wall easily, and for most of the day I found a nice rhythm, losing myself in the details of the stones.

Then, as the two ends of my repair got closer together, I realised that working between fixed points, you will eventually need a stone that fits the remaining space *exactly*. Swiping at the edges of a sandstone slab with the hammer Luke had given me that morning, I clumsily tried to remove the lumps and bumps that prevented a perfect fit. The stone broke perfectly. In two.

By late afternoon, shoulder and arm gelatinous, a drystone wall had appeared in front of me. I ran my eyes over its surface, then placed the cope stones, making sure they were straight. Shouting for Luke's attention, I waved him over and tried to play it cool.

Arms crossed, a frown between two serious eyebrows, Luke

surveyed my work. He walked towards it, and I thought I heard him sigh. Silence. Growing concerned, I searched the wall for any obvious mistakes, though now I wasn't really convinced I knew what I was looking for.

I glanced at Luke, unsure what to say, and saw the smile dance around his eyes.

'You're a fucking arsehole, by the way,' I said, laughing.

'How's it feel to have done your first official bit of walling then?'

Taking a moment to contemplate his question, I looked at my work, grinned, and surprised myself with sincerity. Good. It felt really good.

In bed that night, holding my phone, lower back aching, hands and arms weakened from the tough manual labour, I swiped through photos from the day. Where there once was a gap, now stood a wall. I slid my thumb back and forth. Gap, then wall. Gap, wall. Built by me. Although the walls I created in the garden had led me here, they had eventually been dismantled, to my exaggerated moans of distress, for grass the kids could play on. This wall was the first proper thing I had ever constructed.

Feeling proud, and some kind of powerful, I was aware of a shift within me. In drystone, the finished product is directly related to physicality. There's no separating the two. Looking at what I'd built, I knew without any doubt that my body had brought it into existence. This wall wasn't just my first contribution to the long tradition of drystone in Scotland. No. It was something indisputable. A demonstration of the undeniable value of my physical form. Put simply, when I realised I could build a wall that would last hundreds of years, men wanting to fuck me felt a little irrelevant.

•

After ten years of trying to make it work, Janek and I ended our relationship but remained living together. Although we'd made that decision for good reasons (the kids and money), it didn't make the initial adjustments any easier. We'd been sleeping apart for a while, but now I had the entire post-kid-bedtime evenings to myself. Definitely bored, maybe lonely, I escaped to the vast world of online dating. This time, people did warn me not to date, or at least not to join Tinder. But of course I didn't listen.

When Janek and I met, online dating hadn't really been a thing. My only exposure to that world had been in 2004, when I'd paid to join *Guardian* Soulmates, the online equivalent of a 'lonely hearts' newspaper column. The man I connected with quite genuinely requested that I get the bus to his house, sit silently in the corner and watch him masturbate. I didn't stick around to find out if he was too cheap for a taxi or if public transport really did it for him.

I left the site soon after, feeling out of my depth among the high percentage of academics.

This time around, I spent weeks fine-tuning my bio. It needed to be the perfect balance of honest and witty so I could attract exactly the right man. The kind of man who might meet me in person and buy me a coffee before unbuckling his belt and checking the bus schedule. A girl can dream.

Within a week of advertising myself under the title, 'Probably funnier than you', I'd received all sorts of requests. I'd been taken aback when a man told me he had access to drugs that would paralyse me from the neck down. A tad sceptical when another reassured me, too many times, that if I were tied to a bed with a bag over my head, it would *definitely* be him, and no one else, fucking me. After three long weeks, I was done, only sticking around to see what other nonsense an anonymous person with a penis could come up with. Then I met Mark.

I'd avoided matching with him because in one of his photos he was on a studio floor, shirtless, an unironic, unironic Derek

Zoolander. Now, I swiped right. I sent his profile to a friend who surprised me by saying that she knew him, and so did Luke.

I messaged Luke. He typed his response for a while, three dots blinking in sequential order. His reply, when it finally came, tactfully told me not to bother because, based on what he knew, Mark would absolutely fucking hate my guts. *Don't threaten me with a good time*, I thought and introduced myself to Mark, saying I had it on good authority that he would absolutely fucking hate my guts.

Mark wasn't the bimbo I had imagined. Other men I'd matched with seemed incapable of conversation, but he kept me interested. And, looking back now, I see that what he actually kept me was hooked. The signs were there. Two days of silence after I didn't send him a photograph of my tits; refusals to consider my perspectives on anything. My first taste of freedom, and I had wandered back into the dynamic I'd been working to remove from my life.

We chatted for weeks before agreeing to meet, and I was nervous to take that next step. Going on a date for the first time in over ten years. I might even, finally, have the capacity to connect with someone who was emotionally available. Someone who liked me for me. *And maybe this time I'll let them.* This thought travelled from my brain to my stomach, settling as either anticipation or anxiety, and I got off the bus two stops early to collect myself. It was warm, and this far into Edinburgh's Southside the broad pavements were quiet.

I checked the time, saw it was still far too early and, ignoring Google Maps' protests, took a detour to walk by the Commonwealth Pool. A view of Arthur's Seat appeared between buildings. Waiting at traffic lights, I found myself beside a small beer garden. The clink of glasses and drunken laughter reminded me that this was the first time in my life I was going on a date sober. *Shit.* But then again, I felt stronger than I ever had.

When I saw Southpour ahead, I ducked into the nearest doorway. Smoothing my clothes and checking my hair in my phone camera, I took one last deep breath before stepping back onto the street. I approached the bar, and through its large glass windows glinting in the evening sun saw hundreds of jewel-toned bottles of booze and my date sitting in a booth. Standing up straighter, habitually sucking in my belly, I pushed open the door and walked inside.

Mark stood to greet me and the first thing I noticed was his height. Not six foot one: instead, much closer to my own five foot nine. Smiling my greeting, I busied myself taking off my jacket and tried to hide my disappointment in the careful folding of fabric. I didn't care about his height. I cared about his lie. He had seemed like a confident and deliberately honest person. Unsure what to do with this new information, I put it aside and sat down.

I opened my bag, and with purse in hand asked him if he wanted a drink. He smiled and gestured to the nearly full pint in front of him. At the bar, I studiously ignored the bottles of spirits and their reflections in the mirrored panels behind and ordered a glass of Coke with no ice. Back at the table, I slid into the booth, took a long sip of the dark syrupy liquid and began to talk.

The conversation was easy and interesting. He was flirtatious, and I had forgotten how exciting that skittish repartee could be. I also liked how he looked. Wide-set, almond eyes, full lips and a jawline borrowed from David Gandy. I leaned into it all, into him, and the hours passed pleasantly.

As we got up to leave, I stood close to him and, in the false sense of familiarity that intense flirtation can bring, I put my hand on his chest. Before I could turn a clumsy attempt at teasing into something more graceful, the words had fallen out of my mouth.

'How tall did you say you were again?'

He stepped back and left my hand hanging in the air, then,

picking up his coat, he answered, 'Six foot one.' I felt the shift from the top of my head right down to the depths of my vaginal canal. As we said goodbye on the street outside, his hug was cold and perfunctory, words flat and final.

On the bus after, I felt like I had fallen off a cliff. Endorphins from weeks of online flirtation gone in a second. I sat there, wondering what I had done wrong, waiting to hit the ground. On the train, I checked my phone constantly and, eventually, feeling bothered in a way I hadn't in years, messaged him to say that I sensed something was off and that I'd rather hear the truth than be strung along. His reply came quickly.

I usually date women who are much better-looking than you. I thought your personality would make up for that, but it didn't.

Hand over my heart, slumped in my seat, my life flashed before my eyes. Here was the impact, and the pain would soon follow.

Clutching my phone in both hands, I stared at the screen, now dark, and saw the flush of humiliation on my cheeks reflected back at me. But behind that, something began to solidify. Punitive, insecure, manipulative, his words could have been taken from a pick-up artist subreddit. And they didn't matter.

Although my ego was bruised, his words still echoing in my mind, I knew this was as far as I would ever let them travel. I would not allow them to settle, to become part of me. What's more, I had been strong enough to go on a date sober, in a bar. I was strong enough to read the message again. I did.

And then I screwed up my face in disbelief. Mark had made the mistake of taking the insult too far. Unattractive, sure, we all have our preferences. But unattractive *and* boring? My laughter was a shock to me and to the woman holding a houseplant on the seat next to her. Because there is one thing I know, after everything and for ever. My chat? My chat is immaculate.

•

It was made casually, but my friend's comment about my not-so-neurotypicality stuck with me like a thorn in my side. I hated that someone else thought there was something wrong with me. Despite knowing, with absolute certainty, that she was wrong, she didn't even know me and who did she even think she was, sometimes I wondered what she could have meant.

I knew about ASD, because Janek had been diagnosed a few years earlier, and that didn't seem to fit for me. I didn't know much about ADHD, but it didn't fit either. I was anything but hyper.

I wasn't about to waste real time on this, but still, I read the Wikipedia article. ADHD had something to do with dopamine. Which is implicated in addiction. I learned that addiction issues were far more common in people with this disorder – roughly three times more than in neurotypical people. Of course, this could be a coincidence . . . but there were a lot of coincidences. Descriptions of ADHD began to read like well-crafted horoscopes. I recognised myself everywhere.

Attention, Aquarius! Today, you may feel restless and easily bored, craving constant stimulation and excitement. Enjoy any hyperfocus while it lasts, but rest assured it will not be on the task you actually need to complete. Procrastination is firmly in your wheelhouse. Expect new hobbies.

You may overthink endlessly and be aware of every kind of injustice around you. This could be distracting, so be mindful of finding balance and structure to navigate through the day with clarity and purpose. Embrace your creativity and adaptability and try to ignore your raging imposter syndrome and the belief that those around you will ultimately reject you for who you are at your core.

In the weeks after my official ADHD assessment, I experienced life from somewhere deep inside my own mind. Voices slightly muffled, lights less bright, the outside world harder to reach. I'd strongly suspected that I had it, but having a psychiatrist's confirmation gave me a whole new perspective.

Although I sensed a certain tranquillity, it felt a little too like the calm before a storm.

What I was experiencing was in fact the slow processing of an overworked computer. My mind was taking stock, looking back over the last three and a half decades and reframing everything through the lens of neurodiversity. The heaviness I felt, an accumulation of grief.

A lifetime of failure, assuming that any struggle I faced was due to inherent character flaws. Laziness, selfishness and entitlement. Living this way had killed my ability to be kind to myself and to expect any real kindness from others. How much damage had been caused by not understanding – and not being understood?

Not unlike the early weeks of sobriety, I was overcome by anguish. I had no idea how to reconcile the fact that, had I been diagnosed as a child, my life would be unrecognisable. I would have finished uni. I wouldn't have spent my life fixating on asshole men and what they thought of me. I would have bought food, not spirits, with the Tesco vouchers my mum had sent me that time in Aberdeen.

I grieved for what could have been, and for *who* I could have been. I grieved for the things I did and the things I didn't do. My younger self and the hell she'd been through. Intense guilt, yet again, for what I'd caused my children to suffer.

As long as I could remember, ADHD (and its bff, trauma) had ruled my life. It had taken many forms. I'd found common smells and noises unbearable, spent endless hours alone ruminating and kept my soft toys in bin bags next to my bed in case there was a fire. I had been acutely aware of unfairness everywhere, and this, more than my difficulty paying attention, had made school nearly impossible.

I'd argued with teachers and got into angry debates with other pupils. I recalled choosing the bombings of Hiroshima and Nagasaki as my topic for an English persuasive essay. For the same essay, the boy who shared my desk wrote about why dogs make better pets than cats.

Aged twelve, fixated on, and consumed by, the atrocities, I'd needed my peers to be, too. For weeks, I'd debated nuclear disarmament with anyone who would engage. And if they didn't agree, I'd cry in frustration, my heart aching for all the iniquity in the world. Once, a teacher had to ask me to leave her classroom because I was so enraged. Unwilling, but also unable, to let it go. This may be why I was voted fourth most annoying in my year.

When I saw injustice, nothing else mattered, and where others felt fear, I did not. I regularly stood up to bullies, to teachers: then to bosses and doctors, police officers and librarians, bank officials, snooty store clerks, midwives. When people told me I was brave, I accepted the compliment. The truth was, I couldn't control my reactions. The minute I sensed injustice, it was as if I were compelled by some higher, much angrier power. And I was congenitally immune to hierarchy. Then there were friendships. And work. All I saw was one disaster after another.

For as long as I could remember, I had strongly suspected that I must be better than the mess I actually was. The secret, or so the world had told me, was to try harder.

Even when I failed again and again and again, the party line was that I just wasn't applying myself. And how could I ask for help when I clearly wasn't doing enough on my own? The problems were not really problems, they were ordinary things that other people did all the time and should and could be overcome by my efforts alone.

Setting these impossible standards, and never succeeding, kept me in a constant negative feedback loop. Living life as a fuck-up, I became afraid to fail. To avoid the brutal recurring pain of having my inadequacy confirmed, I either didn't try at all or wouldn't stop trying too much.

Managing undiagnosed neurodiversity in a world that expects neurotypical compliance is impossible. Despite moments of brilliance, intuition, hilarity, you are bound to defeat, to disappointment (your own and others') and to degradation. Unfulfilled potential gnaws away at you, a lot like the longing you feel in unrequited love. All that possibility and no idea what to do with it.

But knowing why I was struggling changed things. The diagnosis handed me so much of the understanding I had been searching for, and it presented me with a neat little label with which others, if they chose, could understand me too. Once the grief had subsided, I attained a whole new level of self-acceptance.

•

Drystone is a male-dominated profession; therefore, the attitudes at the core of the craft have been shaped by men. There's a lot of harder, better, faster, stronger, fuck your knees, no PPE, leave the sunscreen at home, codeine through the back pain, 'dinnae be a pussy'. I know a guy whose favourite saying is 'Life's hard, if you weaken', and instead of wearing goggles, he has his corneas scraped. Even though I recognised the ridiculousness of these postures, it was hard not to fall prey to them.

Initially, feeling like I had something to prove, worried I wasn't strong enough, tough enough, I complied with drystone's machismo. Embarrassed to wear knee pads when working with others, or to take time out when my back was twinging, I tentatively walked the line. The discomfort of kneeling on stone splinters paled in comparison to the ache of knowing that I was once again prioritising the expectations of men.

After a particularly gruelling week, and a weekend icing swollen, bruised knees, I decided to make a conscious choice not to engage in self-flagellation. Every day, I would remind myself that my body was for me, and I was for my body. And apparently, my body had opinions. Tired, sore, thirsty, it would let me know what I needed.

This might sound basic, but to me it had not been. In the past I had paid attention to my body as it neared a crisis point, responding only after it had shut down. Headaches, injuries, feeling faint, a breakdown. Another breakdown. Now that I was listening, I found I could prevent these crises. Responding to my physical needs in real time felt strangely empowering, alluringly adult, and I liked that feeling a lot.

I had been afraid that, being less physically strong than the men, I wouldn't be accepted. But working alongside them, I saw that for every stone I could not lift, there were stones that they could not lift. Some men spent a whole day in the digger shifting huge foundations; sometimes three men lifted one stone together.

We were all working to the best of our abilities. I learned how to build platforms with smaller stones that incrementally increased in height as I slowly lifted the weight of a boulder closer to its position in the wall. There were workarounds but no shortcuts, and although it might take me longer to build my section, I would get it done all the same.

•

Even at thirty miles per hour, the road to Glen Lyon, with its sharp corners and abrupt stops for passing places, has all the intensity of a rally drive.

Checking my phone, I saw that I had no signal. *Fuck*. If Cal texted to rearrange our meeting, I wouldn't know. Then our arrival would become an intrusion. Not the first impression I wanted to make.

It's difficult to know how many drystone wallers exist worldwide. In many countries, as it once was in Scotland, farmers and land workers learn drystone to maintain existing structures that are still of use to them. In the UK, the field is now more specialised. But there are still many drystone wallers who work outwith formal structures, appearing on no official lists. Even if you count them all, the worldwide drystone community is still exceedingly small. The number of those wallers who are online, and contactable, is tiny.

I'd found Cal on Instagram. A man of few words, his posts had been infrequent and to the point. Deciding my next steps, I'd broadened the scope of my investigations and discovered his website: another exercise in brevity. Then, in his 'About' section, between all the pragmatism and skilful work, were glimpses of a fascination and wonder towards stone that mirrored mine.

I wasn't really sure what I'd been looking for. Maybe reassurance, probably advice. Our initial messages had been polite. Him cagey as he tried to ascertain my motives. Me cautious, understanding that he was a Scottish man who had been working all his life hitting hard things with big hammers. We hadn't seemed likely to find much common ground. He owned a digger, and I still got eels inside me at the feeling of wet mud.

Despite these differences, I had felt at ease in his (virtual) presence. From our first phone call, Cal had approached the art of slagging me off with great confidence and gusto. And in his deep voice, draped in a quintessential West Coast accent, it was nothing but endearing.

Looking for praise, I'd sent him photos of my best walls, only the ones I was really proud of. He had sent them back with my mistakes marked up in red. So I had found pictures he'd posted online and done the same thing to him. It was an easy back-and-forth, a groove that felt well worn. Six months later, on this wet April morning, Luke and I were driving to meet him for the first time.

To enter Glen Lyon feels like the discovery of a secret. As if the innocuous left turn half a mile past Fortingall is in fact a portal to a lost, ancient realm. Beneath a tree-shrouded cliff face, a long drop into dappled light on water, where slow-moving pools are black in the shadows, amber in the sun. Around you, drooping branches meet those outstretched, to create something cave-like, the winding road a passageway.

Then, two miles in, the river rises from its ravine to meet the road. Rocks and trees fall away, and a new world unfurls. Bound by vigilant hills, the broad fields of Glen Lyon are home to sheep, drystone and lichen-covered trees. Even though it is sprawling, thirty-two miles in length, it is enclosed. A place all of its own.

With no phone signal, I relied on Cal's directions: a series of hastily written WhatsApp messages. Starting strong, he'd grown bored after 'Turn left at Bridge of Balgie', and I could sense his irritation when I read, 'You'll know where it is when you see the digger.' So, driving the road towards Ben Lawers, we peered into the low light, hoping Cal's instructions were enough to get us there. Pulling my sleeve over my hand, I wiped the windshield and shouted excitedly, 'I see them! . . . I think . . .'

Slowing as he squinted through my window, Luke waited for a road to appear and then made the turn. We announced our arrival over the potholes and stones on the track, heads turning as we approached, and, lifting my hand in the air, I extended a silent, awkward greeting.

Once I'd climbed out of the van, I could see that the bridge was still standing. Old, arched and made from large blocks of weathered grey stone, the dilapidated structure spanned a gulley between the house and adjoining fields. Cal's job was to rebuild it. From scratch.

We walked up the hill towards the digger, busy swinging and scooping on the bridge, Cal at the controls. As we waited near the edge of the stonework, a man appeared beside us. Smiling, dimpled-cheeked, his blue eyes vibrant beneath black hair, he had the air of someone who might refer to themselves as a 'lovely big so and so'.

'Cal?' he shouted over the rumble of engines and metal scraping on stone.

Another man of few words. I nodded.

He waved his hands until he caught Cal's attention, then gestured towards us. The noise stopped. Cal stepped out of the cab and walked towards us, taking off his baseball cap and wiping the sweat from his head before replacing it.

With a lopsided smile, he said, 'You made it, then,' and as I offered a handshake, he extended both arms in the air. For a split second, I thought of my father, our awkward greeting twenty years before.

I said, 'Oh!' before lowering my hand and accepting the hug.

Beside us, the dimple-cheeked man introduced himself to Luke. 'Scott. Nice to meet yous. Just up for the day?'

Luke sprang into social action, shaking Scott's hand and introducing himself before nodding his head in my direction. 'This is Kristie.'

I smiled back.

Scott's face lit up in recognition. 'Ah, yeah, I've seen you on Instagram. Husband and wife wallers!'

I interjected. 'Um, ex-husband, ex-wife. Divorced for ten years.'

Scott looked confused. 'But you didn't change your name back?'

'I kept it. Because it's *sexy*.'

Another awkward pause. Luke, tired of my bullshit, put his hand to his forehead. I smiled.

The tour began. The huge wooden formwork hand-built by Cal himself, the place in the bridge where the cow put its leg through, the neat lines of stone already dismantled. We were lucky we had come up early as the demolition would be done by hand later that day, with the rebuild to begin tomorrow.

As the tour drew to a close, I excused myself and walked down to the nearby cottages to use the bathroom. Turning round at the wooden door, I looked at the three men standing on the hill and was overcome by a powerful intuition. As I washed my hands in the cold water at the mud-splattered sink, my jaw was tightening. Heading back, I heard laughter and then, as I approached, Luke asking Cal about his digger.

I rejoined the group. 'So which one of you said it?'

I hadn't planned to do this.

Making eye contact with both of his companions, Cal spoke for everyone. 'Which one of us said what?'

'Whatever sexist thing that was said.'

Met with silence, I waited a few seconds before continuing. 'Well . . .' I looked around the group, letting my gut guide me. 'I think it was Scott' – I paused for effect, enjoying their incredulity – 'and I think it was something like "the old ball and chain", or "does she always do all the talking for you?" or . . .'

Wide eyes and gaping mouths told me I'd hit the nail on the head. All three looked at me in disbelief. Cal, laughing, looked at Scott, now not just blue-eyed but also bright red and speechless, and said, 'She's nailed you to the wall, man.'

•

If women marry men like their fathers, the closest thing to my entirely absent father was an emotionally absent man. And I married him, twice.

Luke had been a somewhat oblivious father. Through the same permissive parenting he knew from childhood, he had created problems that I then had to remedy. He had supported no routines, refused to make his life child-friendly and stubbornly resisted any and all suggestions. Janek, ever present in physical form, defaulted to the deeply authoritarian parenting he had encountered as a child. Oppressive rules, unreasonable expectations, punishment. They both struggled to recognise their own (and therefore others') needs, driving me, and my daughters, to madness.

For years, Luke and Janek had been protesting, aggrieved that they had not been blessed with the innate skills that I, a woman,(TM) possessed. This stance was hard to dispute. I really was better at most parenting things, and their unfaltering commitment to failure had made the argument impenetrable.

But nothing had come easily to me as a mother. To an outsider, I was a nurturing, affectionate and composed parent who adequately understood and responded to the needs of her children. What an outsider didn't see was the internal battle. How hard I worked daily, minute to minute: to nurture when my brain wanted to kill me, to be affectionate when I don't like most touch, to be patient (patient!) while desperately trying to get things done and held hostage by the narrative meandering of a four-year-old. I am terrible at admin and lose most things the minute I put them down. But I had learned to fill in school forms and make payments immediately, slotting paperwork back into bags, sealed safely in envelopes. Things I mustn't forget placed right before the door. I had become a woman of lists, timers, alarms, pill boxes, stress, guilt, dread. Yet my ex-husbands were offended by feedback on their 'you get what you get' packed lunches and inability to remember that children also need shampoo.

The responsibilities that Luke and Janek cast aside had landed squarely at my feet and, knowing that these things simply needed to be done, I had (almost always) found a way. I never begrudged my children the help they required, but I did feel trapped. Not by what they needed, but by the fact that I was the only one equipped to give it to them.

And it wasn't just aggressively repacking the fucking lunches. I had devoted myself to our children. I had made the sacrifices, I had done the learning, put in the time. When Nell asked for Flower Linda and Strawberry Miller, I knew which dolls to get. When Jo said she'd gone from Radiant to Immortal in Valorant, I knew to commiserate. The kind of socks Nell could tolerate and where to source the preferred gluten-free, dairy-free snacks. The fact that Jo needed Alexa to shut the fuck up so she could think, and Nell's innate drive to explain. To ease spider-based anxiety at bedtime, I'd researched anti-spider smells and bought the essential oils for the all-natural spider repellent that Nell happily sprayed before sleep. Magnesium lotion foot massage for the child who ruminates. And I was tired.

Late one night, researching a sensory-issue-suitable art smock for Nell, thirty-two tabs crowding the top of my small screen, I despaired. Having used up all customary courses of action (nagging, shouting, ultimatums, shouting ultimatums, begging – everything that, as ex-wife, was still in my repertoire), I didn't know what to do.

I closed my laptop. Was I really supposed to do everything alone for ever? And then DIE?

My righteous anger could have powered a locomotive. To run right over my ex-husbands. Who had been tied to the track, mouths stuffed full of school-trip permission slips.

A knock at the door alerted me to Janek, laundry basket in arms. Without a word, he dropped it and retreated downstairs. Through its white mesh exterior I glimpsed pink and blue. Clearly not my clothes. But what else should I expect from Janek, repeat offender. He was banking on me getting up from my bed, picking the items out of the basket, walking downstairs (whence he had come!), going down the hall to the kids' rooms, climbing the other stairs, and putting the pastel pyjamas properly into Nell's dresser. They had been smuggled here. I was again an illicit, unwilling clothes mule.

As my right eye began to twitch, inspiration struck.

If necessity is the mother of invention, desperation is the father of ingenuity. And there were two of them. And I had an idea.

•

On my knees, jeans turning green in the grass, I wrapped my fingers around another long willowherb root, grunting as I pulled it out. Our neighbour, bike at her side, red hair sticking out from her helmet, was waving to me at the gate.

NOFX in my ears (reminding me that I could be truly free, like a shopping cart), I waved hello and waited for her to move away. She didn't. I took a long time wiping off dirt and walking over.

'How are you, Margaret?' I asked, earphones in hand.

'The usual,' she said, embittered but smiling. 'Yourself?'

'The usual,' I replied, shrugging.

Without missing a beat, she launched into a tale about a recent appointment with the gynaecologist that had ended with him telling her that she had the 'lady bits' (her words) of a woman half her age.

After appropriately congratulating the juvenescence of my neighbour's vagina, I began to back away. This felt like a good point to wind things down. But Margaret kept talking.

'Luke's van has been in the drive a lot. I hope everything's OK?' Her face did its best to show concern.

I'd been waiting for someone to bring this up.

'Well,' I said, 'it's been in the drive a lot because he lives here now.' Matter-of-factly, patiently, as one would explain puberty to a child.

'Oh,' she said, as if surprised. 'So Janek has moved out?'

'Nope. Janek's still here.' But she knew that. She'd been chatting with him in the garden two days earlier.

Margaret unclipped her helmet, hung it on the handlebars and leaned her bike against the wall next to the gate. Invested.

'So you're all living . . .?'

'Yep.' I wasn't going to make this easy for her.

But Janek and Luke had made it easy for me to convince them that all five of us should live together. They had each confessed that they were struggling and overwhelmed with the responsibilities of work, money and their roles as fathers, and although hearing that had made me want to tear my hair out, I'd figured that being on the cusp of triumph was the wrong time to argue.

Margaret fumbled for the right question. The question that would give her the information she wanted without requiring her to ask if I was fucking not one, but two, ex-husbands.

'But you're still separated, from Janek – I mean, from Luke and Janek? Are you all . . . together?'

'Margaret,' I said, finally (almost) ready to put her out of her misery, 'if you want to know if I'm shagging both of them, just ask.'

Her face reddened. She swallowed the words, almost gasping. 'Um, I . . . Well . . . are you?'

I paused. 'No.'

The truth: it was an arrangement born of stark pragmatism. A collaboration that sought to conquer the relentlessness of adult life. Not two dicks at once.

'Well, good for you,' she said, visibly deflated, putting on her helmet. And, as if to convince herself, she said it again as she lifted her bike away from the wall. And again as she put her left foot on the pedal. And as she said goodbye, her expression was so flat, so disappointed, that for a moment I felt I should apologise for not being more interesting. Or horny.

Back in the soil, I thought with appreciation of the handsome exes everyone hoped I was fucking. Men who had been willing to see beyond their own egos to do what many would not. To create something better, more balanced. And that 'something better' had extended to our children. Now my daughters saw their dads every day, and those dads had become less stressed, better informed of their kids' hopes, dreams, fears – and lunch preferences. Two men do not equal one woman. But they would do.

It wasn't idyllic. Far from it. We still fell out, shouted and jostled to have our needs met. Our ideas clashed, our opinions could be at odds, but we were working it out. We managed family dinners every evening and, on Sundays, because everyone loves Nell, a board game night at her request. It may have piqued our neighbours' curiosities, but our home functioned roughly how any home would function. Except now there were three parents and only one of them was me.

Setting the Throughs

Balancing on the partial cheekend, I put my hands on my hips and look down the length of my work. Thirty metres of wall, all built to through height. Roughly up to my knee. Halfway there. I jump from the cheekend and pick up the first through stone from the pile. It's a neat oblong of flat sandstone, and I position it across the two sides of the wall, wiggling it against some protruding hearting, letting it settle into place.

Walls can of course be built without, and sometimes they have to be, but through stones add an extra layer of support. They bridge the width of the structure, and by way of drystone magic, bind it, helping the wall to settle as a whole, rather than as individual parts. Stones of this length are hard to come by in just two loads of fieldstone. In a thirty-metre wall, there should ideally be twenty throughs. Sometimes there are none to be had; this time I have eighteen.

Next in the pile is a bulging, uneven stone that, with some work, might be called a triangular prism. My eyes skip over it and I remind myself that it has to go into the wall sometime. It might as well be now. I place it, reach into the bucket behind me, grab a handful of hearting, and begin to pin the stone.

These mid-point structural additions increase longevity by taking on some of the unknowable burden the wall will experience over time. They future-proof my work.

Until the end of August, fieldstone had been friendly enough. Two weeks of relentless autumn rain had changed that, transforming the stone's dusty surface to a slippery, wilful nightmare. It was either warm enough for the wet mud to be so thick that simple movements were gruelling – or cold enough that our first task of the day was to ruthlessly sledgehammer apart the huddle of frozen stones that had sheltered together overnight.

The stone here had not been hand-picked. There were stones that only a mother could love, stones that might not even be stones, and stones that were good stones but still the wrong stones for this wall. Even so, amid the mire, it was our job to find the right stones and the right places. Trapped fingers, tired arms, cold faces, wet clothes, sore knees. Luke, responsible for the completion of the project, mostly left me alone to work on my section of building. Miserable. Days passed where I put ten stones in the wall.

I've never been good at being bad at things. I become obsessed, get proficient, realise I'm not the best the world has ever seen, then consign it to the ADHD hobby graveyard. I'd worried it would be mundane, that drystone was going to go the way of paper cutting, Japanese pull saws, embroidery, cycling and foraging. Driving home after work in the dark, muscles aching, ego bruised, I'd curse the craft and tell myself I was done with it. So it was a surprise each time I found myself back at the wall. Yes, drystone had got under my skin, but also, I'm stubborn. I couldn't let them win. Not the stones, not the men.

Drystone has always been male-dominated, and its culture is heavily influenced by that. You've only done enough if you end your days lying prostrate, the Gods of Walling judging you from on high. 'Part-time waller' might as well be a slur. Attempts to open minds are warded off with protective charms or vehement proclamations of being unable to see race. Any suggestions for better working conditions go over about as well as Oliver Twist seeking organic, free-range butter for his gruel.

So it's maybe no surprise that the craft has made little space for the needs of women.

Once you've taken some beginners' courses (or taught yourself the basics), you should find someone experienced to learn from. In Scotland that means you will pick, almost exclusively, from male wallers. First, you have to locate him. Then persuade him to take you on. If you are successful, you will go and work with him, in a field, in the middle of nowhere. Even within the professional register, there are no background checks, no way to ascertain what a man is like before you head out to meet him and lose mobile reception in the middle of nowhere. If Luke had not existed, this probably would have been the point I decided that walling wasn't for me. #NotAllMen but #AnAwfulLotOfThem.

•

I was taken in by Brian's praise. Garrulous and opinionated, he had the confidence of a man who would describe himself as 'intelligentsia'. (He did. He described himself as 'intelligentsia'.) He told me that working with him was a great opportunity. That he was influential in the industry. That association with him would raise the profile of Luke's business and we could learn from someone who knew a lot more than we did.

But in the early weeks of messaging, I already felt uneasy. When I got a notification from Brian, I'd hesitate before opening it. I wanted to be friendly, but he was too quick to turn the conversation to relationships, and had already described the sort of lover he was (a very good one). Over the course of my life, I'd met lots of men like Brian, dealt with many messages like this, many times before. I didn't want to fuck things up for Luke. I was just being oversensitive.

Luke may have hired him, but it was me who found ways to manage him. Deflecting his compliments about my shoulders and his looks towards my breasts by finding my fleece and zipping it up (even on hot days). Vaguely nodding to his jokes about sex, ignoring remarks on how it's a man's job to satisfy a woman in the bedroom, giving no attention to conversations about incest, turning my back to declarations of how beautiful his dog's arsehole was (from a craftsman's point of view) and refusing to take the hint when he held out a bottle of sunscreen, saying that he had been too busy to apply it to his (naked) torso. He was worried about cancer.

Sorry about that, pal.

•

I had expected to find Jo at her desk, shouting objections at Fortnite teammates. Instead, she was on the floor, surrounded by ragged-edged pieces of paper and coloured pens. When she saw me in the doorway, she waved me over.

She pointed at her screen, bright in the dim light of evening, and started to speak quickly, angrily.

Thrusting three pieces of paper towards me, she said, 'I made a survey online and so many people answered, like most of my year and a bunch from other years too . . .'

Taking the sheets, I saw a list of questions, some scribbled out, some underlined. But before I could read, she was pointing at the screen again. 'There's little pie charts showing how many people said what.'

I read aloud. 'Seventy-seven per cent have witnessed racism at Crieff High School. Fifty-five point two per cent have' – I ran my finger across the screen – 'experienced sexism.'

I'd found myself shocked by the things Jo had told me since the Night of Champions. That white kids would shout the N-word down the hall, that Blair (the only out trans kid) would be viciously taunted at lunch. And, of course, that Jo had continued to be called a 'Paki'.

Scrolling again, Jo said, 'Over a hundred people responded. Fourteen per cent of the school. Right?'

Checking her arithmetic on my phone, I confirmed with a nod. I sat on her bed, silent for a moment, trying to take it all in. 'You didn't tell me you were working on this.'

Jo looked worried. 'Did I do something wrong?'

Her head was restless on my chest when I leaned down to hug her. 'Absolutely not. I . . . you put all this together yourself?'

'Yeah . . . You know they told Blair to kill herself? It was the last fucking straw. And look' – she scrolled down – 'I put in a section for comments, and it's not just Blair. There's over ninety responses. I'm so angry, Mum.'

On the full page of text, separated by bullet points, one line immediately stood out to me. 'What,' I asked, 'is Slap Ass Month?'

'A month where the boys in fifth year slapped as many girls' asses as they could.'

I kept reading. 'Explicit photos of an underage girl shared around.'

'It got so bad she had to leave school.'

I tried to tear my eyes away. I couldn't. 'Groups chanting "Kill the Pakis" in front of teachers, pupils threatening to out LGBTQ+ students, sexual assault and harassment, students of colour called the N-word or "monkeys" . . .' I felt tears welling up. 'Gay student followed home and threatened with being rolled up in a carpet and set on fire.'

These stories were from children, some as young as eleven. 'Jesus fucking Christ, Jo. It's like this every day?'

With the same long eyelashes and warm brown eyes as my mother, Jo looked at me and told me the answer I already knew.

•

Crieff High is a typical rural Scottish school: although its catchment is sprawling, it has little diversity. I'd gained the impression that attitudes (and policies) came from an assumption that, because there were so few people to be bigoted towards, bigotry need not be considered. I'd sent dozens of emails, spoken to all members of senior staff on several occasions and been to three in-person meetings. I was frustrated that they hadn't taken adequate action, and they were even less interested in hearing that the action they had taken had not been adequate.

Like the school in Edinburgh, they seemed to expect Jo to be more flexible, factor a bit more bigotry into her day. And I was on the 'any bigotry is too much bigotry and if you can't see that I don't know what to tell you' side of things. Jo too had got a few teachers to listen, but, like us, they always fell at the final, senior-management-sized hurdle.

Three days after her survey, Jo shared an open letter online. She asked me to share it too, and I did, unaware of how far it would travel. Late morning, my phone rang. Yes, I was Kristie De Garis. Yes, I was the mother of the girl who had put out the letter. No, I would not like to be interviewed right now. No, I would not put her on the phone right now. In fact, I needed to go right now. I needed to speak to my daughter.

But first I called my mum. I knew she would have an opinion, and if it was a bad one, I wanted her to tell me, not Jo.

The end of her words becoming more guttural, my mum reconnected with her Glaswegian accent. 'She's her nana's granddaughter!' she said proudly. But, she told me, she didn't think I was worried enough. I let that one slide, because I absolutely was.

Later, I explained to Jo that once you speak to a news outlet you have little control over the narrative, that it can be a lot to invite the world into your life. That she should feel no obligation, she had already done enough. Feeling more than a little guilty, I told her this was a job for adults, not children. Jo listened politely. Poised, head cocked to one side, she asked, 'Do you think this will make things better?'

'Oh, I don't know, babe,' I said. 'Maybe?'

Suddenly out there in the world, showing all her cards, my child had become deeply vulnerable. My instinct was to brief her for hours before interviews, interrupt her if she strayed even slightly from the questions she was asked. This could all go so very, very wrong. But I held back, as much as maternally possible, and carefully offered general guidance. After all, it was her courage, her desire for change, that had brought us here.

Two whirlwind weeks passed, as close to the impetuous, fickle world of politics as I will ever get. Perth and Kinross Council released a statement. We were 'invited' to a lot of meetings. At our last one with the headteacher, I held Jo's hand while he lost his cool so spectacularly that I, like my own teachers so many years before, had to ask him to take a break to calm down. His Rumpelstiltskin shoes squeaked as he saw us out into the hall. We waited.

Twenty minutes later, having again 'invited' us back in the room, the headteacher was defensive and domineering, repeatedly telling Jo what he had planned for Pride month while she asked him to *just listen*. Instead, he had us escorted from the building. A flustered staff member walked behind, ushering us down the stairs. Jo cried and asked if they were going to kick her out of school. I fought, and conquered, the urge to kick over a bin.

After that, the Council announced that a major inquiry would soon be underway. The media discovered that the man heading it had, until recently, worked in senior management at Crieff. Jo released a second statement saying that actually, no, it isn't appropriate to investigate oneself. I filed my own requests for information and discovered that most of our complaints had never been logged, never mind dealt with, including what I'd told Mr Duncan at the Night of Champions.

Then *The Times* picked up the story. They ran a long piece showing how other pupils had been failed by the school and a weekend segment where Jo was compared, in passing, to Malala Yousafzai and Greta Thunberg. Even comments on social media were at least 60/40 in favour of Jo's letter, with many parents sharing for the first time what their kids had been through.

I juggled all my usual responsibilities with the endless emails, phone calls and interview requests. For entire days, I felt painfully alert, unshelled. I sat in on the biggest interview, silent, in a chair, off to the side, as Jo spoke directly, forcefully, into the camera.

The film crew set up in our kitchen, cameras on stands next to net bags of onions. I'd tried to tidy the house before they arrived but as I watched Jo use the idealism of youth to claim her right to a better world, my eyes fell on the windowsill behind her. Nell's homemade fairy potions, a broken Christmas candle and the lamp with a dirty, Sellotape-repaired shade I'd been meaning to throw out for a year. *Shit.*

But Jo was on fire. She was saying things could be different. She didn't need to know how, or what level of government, or on whom to lay blame. The nation watched, in real time, as my Joanna Banana, just sixteen, spoke change into being, right there, right then. I sat in my chair, off to the side, with my heart breaking open.

I'd thought I had no memories of her first year but then I was picturing her tiny baby toes. The feel of them digging into my side as I nursed her. Jo in my arms, reaching across me for her favourite toy, a plush sunflower, petals worn bare by her fingers. And I remembered her, five, drawing an impossible, imaginary plant. Her idea of bamboo to scare away spiders.

•

I slowly chipped away at a hard sandstone block. Standing opposite me, Brian worked on his own piece of stone. It was unseasonably warm and, keeping my eyes turned down, I hit the chisel a little harder than usual, angry that I had to deal with this again.

Luke and I had discussed Brian's behaviour a week earlier. It was inappropriate to smoke weed and take one's shirt off at work. Luke had spoken to him, requesting that he remain fully clothed and refrain from hot-boxing his car in our client's drive. After some reluctance, Brian had agreed.

But two feet away from me, he was stinking of weed, naked torso and wobbling, ageing pectorals.

Luke approached us with a boulder in his arms, and I hurriedly retrieved chisels from the surface of the banker. As it landed, the large stone's weight shook the wood between us. I continued to work in silence, telepathically relaying to Luke how pissed off I was. When that didn't do the trick, I lifted my foot, out of sight, and kicked him. He kicked me back.

'So how is Jo?' Brian asked, hammer poised in the air, waiting for my response.

I sighed internally. 'Good, yeah, thanks for asking.' I kept my eyes on the stone in front of me.

He continued, hammer still in hand but now resting on the banker. 'I find it so wonderful that you have crafted such a cosy little living situation. I take my hat off to you all, I really do.'

'Cosy' stood out in a way that it normally wouldn't. Grimly certain of where this was going, I replied, 'Thanks.'

With both hands on the wood in front of him, he leaned forward. 'It's a question, isn't it? After two failed marriages, are you tired of cock and ready to make the transition to vagina?' He said the last word like he was teaching me the name of an exotic arachnid.

I hated being back in this space, my body and my reactions under a microscope, subject to paralysing scrutiny. I hated my silence and I hoped every day that Luke would make the call and tell him to pack it in and pack up. Then it wouldn't be me ruining everything.

Back at work after a long weekend, Luke discovered that Brian had come to the site and worked on the project without permission. And the building he'd done was just . . . bad. Uneven. There were unnecessary gaps, stones rattling loose on the top course. Quite honestly, it looked like it had been built by a massive stoner. A stoner who was preoccupied with the perfect puckering of the canine anus.

I watched Luke over his shoulder as he typed the email, struggling to find the words to make his position clear. I made some suggestions. Brian's drug use and poor work were an issue. Brian shouldn't worry about completing the current job. Brian would be paid in full.

He didn't respond. He ignored Luke's emails, then his texts, he lied to the clients, he caused a fuss, and it all became a roaring mess.

I struggled to understand how I had been taken in by him, by his platitudes and his promises. He had made it clear who he was from the start, and after all the work I had done, everything I had experienced, all the therapy and sobriety and confrontation and running up hills and photographing flowers and trying to heal my family and fighting for every single fucking thing, I still hadn't shut him down immediately.

A few weeks after Luke let him go, I learned that – surprise! – Brian had a solid, sordid history of sexual harassment. My thoughts formed a knotted lump in my throat as I considered how much my silence had enabled. With Brian, but beyond him, too. I wondered how many of the other men who had harassed or assaulted me had gone on to do that to others (all of them). Those men with no boundaries, the endless line of women left to carry the burden.

Brian is just one of many. There was another waller who became fixated enough to harass me, a woman he has never met, for long hours of his one precious life. The violation of bullshit legal letters at my front door. Men like this aren't happy until they've pissed their lurid entitlement over everything, especially everything they are no longer entitled to.

Despite my growing file of bullshit legal letters, most male wallers find it impossible to believe that this sort of behaviour really happens, that misogyny is as present as I say it is. They will often suggest that I've just been unlucky. But every single female waller I've spoken to has had similar, if not identical, experiences. Some worse. Most of those women do not speak up, and for good reason.

A traditional craft with traditional attitudes, drystone seems reluctant to accept the truth. Reluctant to condemn men who perpetrate harassment and intimidation. Reluctant to make change.

•

I lay on the floor, a cup of cooling hibiscus tea beside me. I didn't want it. I wanted my fingertips in cold condensation, the crisp smell of Pinot Grigio, the satisfaction of that first sip. Without turning my head, I side-eyed the 'Life's hard if you weaken' mug Cal had given me for Christmas. I would prefer a pint of Black Isle Brewery Yellowhammer, weighty in my hand, straw-coloured in the glass and zingy grapefruit and bitter hops on my tongue. But what I really craved was inebriation. Giving myself over to gravity, mind numb, room spinning. I missed it.

Booze is the fastest getaway car for a brain on the run and nothing else offered the same escape. Exercise required effort. Buying things online was ruined by the very fact of delivery. I had tried eating a tub of triple chocolate cookie dough crème brûlée ice cream, but the impact, on all but my digestive system, had been paltry. The back of my head was getting numb. I sat up and began to pick the carpet's nylon fluff off my sweater as I pondered the liberating qualities of drystone. It was good for me, it made me feel good, but it was hard graft.

It began to rain.

Through the rickety wooden doors of the Tomnah'a polytunnels, the air was humid and herbal with an intense floral finish. In front of me, neon orange twine trailed between voluminous bubble-gum-pink roses. Above them, dill plants with their fireworks of yellow flowers and, fading into the distance, pink and purple pincushion polka dots.

Crouching on the warm, dry dirt, I listened to the downpour: a percussive symphony on the taut plastic sheets above my head. Fuchsia echinacea tangled with delicate daisies, columns of lilac salvia supported by wooden stakes. The humidity and the polytunnel scattered the light better than any of my over-priced photography equipment, making the colours more themselves.

I'd fought the urge to photograph the flowers for two years. I hadn't wanted to be a cliché. Nearing midlife and filling SD cards with intimate close-ups of petal and sepal. And wasn't it plagiarism to simply capture the beauty of something beautiful? Now, in the absence of certain things and in the discovery of others, Tomnah'a Market Garden had become a refuge.

On my first three visits here, I'd rushed in, a slave to my senses, lost my fucking mind and photographed every single thing. Looking over my images at home, outside the fog of aesthetic delirium, all I had seen was wet enthusiasm. No discipline, no composition. Fervour rather than flowers. Dragging two hundred icons to the trash folder, I'd decided that the next time I went to the tunnels, I would go for the experience itself. Nothing more.

I rose and placed my empty camera bags on a sturdy little table near a leaning tower of plant pots. Wandering the space, my eyes didn't stop moving. They settled for a few seconds before being distracted by something else. And something else. A manic feast of floral maximalism.

For thirty minutes I let my cameras hang on my neck. And simply noticed. The coming together of texture, colour, proportion and space. The way the lines of the flowers, doing what they wanted, interacted with the lines of the structural beams, doing what they needed. The orange of a calendula picked up in the rust of a broken spade. Slowly, my brain saw balance within the visual clamour. It quieted, and I got busy taking pictures. To think that a year ago, the equivalent had been two bottles of wine and Joni Mitchell singing about pretty men and pretty lies.

•

The Council's investigation made everyone nervous. Randomly selected students were pulled out of class and asked, in groups, if they had experienced bigotry. A tactic which seemed to me an ingenious recipe for peer pressure and silence. Predictably, the Crieff pupils closed ranks. Rumours circulated (were allowed to circulate) that this was all happening because Jo had falsely accused a boy of rape, that fifth years who slapped non-consenting asses could go to jail.

I was walling out on a back road when I received the final report by email. As usual, I had a shitty signal and the document wouldn't fully load, but what I did see sent me, stone in hand, straight back to the van. I realised too late that I could do nothing with the pig-headed whin in my fist, and it sat heavy in my lap all the way home.

That evening, in a Zoom meeting with the head of education for our area, I pointed out that asking a predominantly white, straight and cisgender student body if they experienced racism, homophobia or transphobia would lead to only one conclusion. Her face fell. She changed the subject. Shoulders were shrugged, questions avoided, and the inquiry had fulfilled its perfunctory obligation. We hung up, I closed my laptop and I talked myself out of getting very, very drunk.

The next day, the BBC's headline was 'NO CULTURE OF DISCRIMINATION AT CRIEFF HIGH'.

Angry tears fell from Jo's face and onto her white school shirt. 'But they're lying, Mum! They're saying I'm a liar!'

My words were weak. 'I know. It's not fair.'

Jo bravely went back to school. But for her sins, students and staff avoided her, and voices shouted 'Grass!' in the corridors. One day after school, squeezed in the front of Luke's van, Jo pulled out her Pokémon sticker-covered phone and played us a muffled recording. Voices, furious shouting. A girl, accusing, and Jo, trying to simultaneously explain herself and get away. She'd been cornered by a mob of angry, misinformed teens, her back against the wall.

Concerned for Jo's safety, and aware that our relationship with the headteacher had completely broken down, we requested mediation. He refused.

As months passed, disillusionment crept in. The more time Jo had to process, the less she trusted any teacher at Crieff. Those who should have protected her, and others, had not. Soon, she stopped going to school and my quiet, opinionated daughter became just quiet. In our late night chats she asked me, again and again, why everyone hated her.

'I was trying,' she said, picking at an already red patch of skin on her knee, 'to make things better.'

I watched my teenage daughter learn adult lessons. *The truth is*, I thought, *most people are fine with things the way they are.*

•

At the top of the hill, the village stopped abruptly. No smooth transition: instead, one last bungalow and a single, sentry lamppost between manicured gardens and open fields. The hedges around me were full and high, dense soundscapes of birdsong and buzzing. There was something about how sounds sat in warm air. Mingling, faded, carrying farther but softer. An early summer, and COVID lockdowns had arrived in Scotland. I was making the most of both.

Sobriety had already painted bold borders in my life: I'd

grieved what others, now that bars, restaurants and social gatherings had become out of bounds, were suddenly losing. The lockdown hadn't made my world smaller. So much had happened in the year before to process and piece together. I was tired. And deeply, primally in need of rest.

Up ahead, a sun-bleached wooden sign pointed me in three directions. I took the left turn and found myself on a thin track, brown and bumpy, winding its way between tall grass and unfurling ferns. As I walked, small puffs of dust appeared around each foot, and the fine particles settled on my sandals, dulling the black straps to a soft grey. A tideline of mud between each of my toes.

As trees appeared at my side, cold air reached out from under their branches, rousing the hairs on my arms. Standing in the shade, I felt grateful and lucky and so at peace.

The threat of social isolation had once scared me, but now I had somewhere else to be.

•

I answered the call, annoyed that Jo would phone me from the other room.

All she said was 'Mum' before the line went dead. I ran to her. She was in her bed, hair soaked with sweat and face so pale I could see veins, ethereal blue-green beneath her skin. Her body shuddered, teeth chattering so badly that I could barely make out what she was trying to say. I yelled for Janek.

When the government had lifted the lockdown, I had resented the intrusion upon my carefully curated life. And over the next year, for reasons based on capitalist production, and not my children's safety, mask mandates were tossed aside, too. Jo had caught COVID from her boyfriend's grandfather on a short, open-windowed car ride. She'd masked in public but it had found her privately.

Although worried that I was just being dramatic, I'd been afraid for Jo. She'd lost her voice and hadn't left her room for weeks. I'd never seen viruses as innocuous visitors. At twenty-one, an encounter with the flu had left me with Hashimoto's thyroiditis, a life-long reminder of a brief and seemingly uneventful time together. And this virus was so new and so high-spirited. Plus, people were still dying while everyone else acted like COVID had transformed into a harmless pathogen.

Here, in the back of our car, I supported Jo's long body, her arms and legs folded in on themselves as she shook uncontrollably. She looked so young and so scared. Watching her with the same terror of fifteen years earlier, I prayed to that God I still didn't believe in, made Him the same promises.

The hospital found no cause for the violent shaking that had lasted hours. We brought her home. Unsatisfied, I called our family doctor, and over the coming months they ran every test they could. Once they found nothing more to investigate, we were largely without medical support.

Each day, I soothed myself with the salve of others' expertise.

In a world so vast, so ultimately unknowable, I deal with my insignificance by trying to know everything. And I like to learn from those who know a lot more than I do. I research the big stuff, like being pregnant, raising children and staying healthy. The small stuff, like buying electrical appliances, choosing paint or where to travel. I've dropped out of university seven times, but I would make an excellent eccentric professor. I read scientific studies and online reviews in equal measure. Whether you authored a paper on SARS-CoV-2 viral persistence or wrote a review for the Sonic Bomb alarm clock, I've read it, and I thank you for your service.

My healthy, sporty sixteen-year-old now slept eighteen hours a day. Even the Sonic Bomb, vibrating in a metal tin of river rocks next to her head, couldn't wake her. I told myself that it was probably residual stress from the incidents at school. But as more months passed, her health further deteriorated. Worried what another COVID infection might do to her (to any of us), we continued to mask, continued to be cautious.

I knew we were doing the right thing, but conversations with friends and family became strained. Again choosing to live in a way that deviated from the norm, I had become a mirror into which others looked and saw judgement. My daughter's illness a reminder of all that they were trying to forget. But after a lifetime of people-pleasing, I knew that I was going to continue doing what I needed to keep my family safe. We would be a sober, platonic-throuple-parented, neurodivergent, anti-racist, anti-sexist, anti-COVID blended household. It felt surprisingly uncomplicated.

•

The thunderclap shook the house. Lying awake, heart thumping, I watched the room light up with the intensity of a crime-scene flash. Another roar quickly followed. The storm was right overhead. In the moments between sonorous crashes, the thick silence, the house sounded different.

On the stairs, I felt my way with fingertips and muscle memory. The porch light shone through the window and onto the floor. The kitchen tiles were shimmering. My brain, somewhere between sleep states, registered that something was unusual, but not what. When I stepped onto the final stair, I was awoken by the shock of cold.

Walking through the water, I could feel mysterious tendrils tangling between my toes. I half expected the water to stop where the kitchen ended and the hall began. When it didn't, the word 'flood' appeared in the corner of my mind. This was a flood. Our home had been flood-ed. A metal lamp, half submerged and flickering on the floor, caught my attention. My body sprang into action. I ran to the fuse box, knees lifting high away from the water, and turned off the electricity.

Luke was still asleep, the water mere centimetres from covering him. I woke him and explained. But he only really understood when he reached for his phone, now floating in its case near his head, anchored by the charging cable extending to the wall.

Janek was next. Startled at the intrusion, he sat bolt upright, put his feet flat on the floor and yelped. The kids were still safe upstairs in rooms that smelled of sleep. I woke them, bright and calm. I asked Janek and Luke to carry the girls downstairs through the hall, the kitchen, water that smelled like wet mud, something sweet, and faeces. Then up the stairs that led to my room. When everyone was settled, I ventured outside.

Dark. Rain. In loosely laced hiking boots, I stepped into the ankle-high water coming down the hill and promptly lost my balance in the force of the flow. As I picked myself up, I gripped my shoes to my feet with tense, curled toes and lifted my sodden pyjamas by the waistband. I waded to the neighbours' doors to alert them and waited for them to answer. Standing in the street, I watched lightning fork across the sky and felt its sonic shockwave in my body.

As the sun rose, the flood waters slowly receded. Every single thing was wet. Even my fingers were wrinkled. I made my way to the shed at the end of our garden, squeezed down past the bikes, sat on an upturned bucket and sobbed. I cried for the cosy spaces I had made, every single thing I had chosen with care. I cried for the loss of my home, a home that told me every day that I would never have to go back to Fairview. And I cried for how hard this was going to be.

Emerging, red-faced and puffy, I found two neighbours in our garden. The same couple who had popped their heads over the fence when we first moved in. Their place hadn't been hit so hard and they were here with trays of bacon rolls and hot tea. Another neighbour was in our kitchen vigorously sweeping the last of the water and mud out the door. They were so generous in the moment we needed it the most, their gestures so kind. I was embarrassed, terrified, by how much it meant to me.

•

After the flood, I had no choice but to be completely present, hyperaware and totally on it. There was something horrific about being so lucid. The insurance company put us in a Travelodge. Kids, adults and every essential belonging crammed into two small rooms, we processed over McDonald's, and at the weekend, roast dinners in polystyrene trays. Every waking minute required a phone call, an online form or a trip back to the flood zone.

Finding a new place to live, managing a renovation, documenting and repurchasing a thousand items, being a mother, working as a photographer and navigating a global pandemic were *at least* five full-time jobs. And there were Janek and Luke watching me from the Travelodge's single beds: wide-eyed, innocent newborns. They may have inhabited the same house, but they never seemed to inhabit the same life.

Nell had left her favourite Supertato book in the living room and asked constantly if it was OK, because she knew it wasn't. Jo wanted to know when we could replace her water-soaked gaming console. And it felt like both asked every hour, on the hour, when we could go home. From across the cluttered room, I watched Luke and Janek shrug their shoulders and tell the kids to 'Ask mum'. My sobriety was being stretched thin.

'Janek,' I said, 'take a break from work, you're dropping me at the house. Luke, make sure the kids get outside and eat something that isn't a burger.'

Stripped back to its frame, beams like bones and wires like tendons, our home looked as flayed as I felt. At the fuse box, I flipped the switch, then, stepping over exposed floor joists, I walked upstairs, plugged in the extension cable and lowered it out the window.

Back in the garden, I grabbed the dangling white plastic, pulled it over to the grass and picked up the hair dryer.

The first box was the size of a small freezer. I tried to drag it over the gravel, but it barely moved and the cardboard began to tear. It was heavy, full of paint tins and offcuts of tiles. I should have thrown these out ages ago. Into the skip. Next, books, some of my favourites, old copies I'd saved since I was young. Speckled prettily and tragically with mildew. Then board games, Trivial Pursuit, Frustration, Monopoly, buckled and bloated. For the hundredth and second time that day, I told myself, *Just things, not people. You can replace them.*

As I heaved the next box off the pile and onto the grass, its bottom bowed and flexed. When I pulled at its cardboard flaps, they lifted and fell noiselessly. I looked inside. The water, given time, had soaked through almost every sheet of paper. If I'd done this earlier, I could have saved them.

Thank fuck the kids weren't there.

I extricated the first clump and began to peel its sheets apart. Five crayon drawings of Pokémon. I laid them on the grass where they seemed to float on tall green blades. The garden had, apparently, profited from the flood water's questionable nutrients. The next sketches were felt pen – or had been. Ink now pooled into blotchy stains at papers' edges, original designs lost for ever. I blow-dried them a little anyway.

Over the whine of the hairdryer, a voice. Margaret was at our gate, bike helmet already off and tucked under her arm as if to mark the solemnity of the occasion. I raised my hand and she shouted, 'If you need anything, let me know, OK?' Taken aback by the sincerity of her tone, I began to decline, telling her we were all good. But she gestured towards the contents of our home on the gravel all around me, said 'Don't be daft!' and continued down the lane.

Next, I rescued a beach scene by Nell which had been stored in a Ziploc bag to stop cotton wool and sand from doing what they do. Below that, a stack of drawings in pencil, stubbornly stuck together. Setting them aside, I picked up the next pile, and the next.

Someone cleared their throat. On the other side of our wall, an older man I didn't recognise. Awful to be one of two houses that had been flooded, he said. Did I need any help? He was old but could still carry things. I thanked him profusely, my voice catching in my throat as I told him, 'It's just organising today.' He nodded and walked on.

Accepting generosity still felt somehow fraudulent.

Back on the grass, I reached into the nearly empty box and found Jo's first attempts at writing. Blurred. One self-portrait I had always loved showed her between a half-eaten sandwich and a crying caterpillar, her full name written above in letter-ish shapes. It had begun to dissolve. I laid it on the grass in the hope that, as it dried, it might find a way to hold itself together. *Just things.* But not replaceable. My hands smelled like the cardboard as I wiped my tears.

Opening the last box, my heart wrenched in my chest before throwing itself from my body, turning around and punching me in the gut. My childhood photos. Sticky and tidemarked.

Their ink, now glue, held them steadfastly together. At the top of one pile, me as a toddler, smelling roses in my nana's garden. She held a lit cigarette in one hand, the back of my dress with the other. On the next clump, a photo of my mum, her shoulder pads, perm, hoop earrings. I tried to pull apart the prints and succeeded only in decapitating my mother.

One picture, mildly water damaged, showed Matthew and I dressed in our Sunday best at the net-curtained front door of our flat in Jedburgh. Breathing disrupted by the convulsions of a determined sob, snot flowing freely from my nose, I thought of how I missed him. How we hadn't been able, after all, to find our way back to each other.

My wet eyes were drawn to a warped envelope. The photos

my dad had shyly given me twenty years before, in Aberdeen. I turned them over, examined his handwriting: too-big loops and strides of smeared blue. Destined for me to lift their small ghostly shapes and place them, carefully and without malice, in a rented skip.

A steady stream of neighbours had passed our gate, commenting on how unlucky we had been. The couple from up the road offered apples and plums from their garden. Others recommended trustworthy contractors or friends in insurance who would 'keep us right'. My pockets filled with scraps of paper, hastily written numbers.

Being offered a bag of apples by a neighbour in a time of need may seem straightforward. But for me it wasn't simple at all.

Generosity had always been a too-bright light that revealed shadows. My need for kindness, and absolute inability to accept it. Pain from the times others were cruel, the times I was cruel to myself. The lingering dark corners of self-hatred and shame; if those offering help discovered the truth of who I was, they would feel cheated. I was taking things meant for someone better.

Just neighbours being neighbours: unbidden, unexpectant. With no idea what it meant to me. They were writing down the names of local builders and I was fighting the urge to grab their hands and faces, to tell them that, through their actions, they had assured my worth. But that would be very fucking weird. So I avoided eye contact and, no longer able, or even trying, to mask my tears, thanked them and accepted everything.

•

Some wallers use only hammers, some prefer chisels. Some stick to a few favoured implements, some have leather rolls full of options. Within our overstuffed tool bag, Luke and I had at least three chisels distinguishable only by the differences in their forging angles or the thickness of their cutting blades. And hammers, so many hammers. Some to hit stone directly, some to hit chisels hitting stones.

As my building progressed from field walls, seen solely by sheep, to garden walls, seen mostly by people (and some sheep), I reluctantly learned to shape stone. A craft within a craft. I was unskilled and clumsy. Learning to shape stone is one of the top five most frustrating experiences of my life.

Each stone type has a preference for how it is approached, and you only glean those predilections through failure. If you hit a stone and it fires back with a shrill 'ping', you are likely dealing with basalt or dolerite. Igneous rocks, tough and not well-suited to shaping. If you try, putting all your strength behind hammer or chisel, the stone will resist, eventually giving up to split in ways, places, you'd never have expected. These dark-coloured, hard rocks are collectively called whinstone, which is hilarious, because you can't.

Sandstone likes to split along its bedding planes, and you can use that to your advantage if you need to take off some height, or create precious shims. Soft sandstone will break like over-toasted toast and crumble under too much pressure. Harder sandstone's weaknesses are better hidden – though it's taxing on the wrists. But slate will always gratify. No bedding planes, instead planes of cleavage, and it does indeed cleave.

Stone prefers decisive action. You don't need to watch a drystone waller to know how experienced they are. You can hear it in the sounds of hammer on chisel, metal on stone. The pace, the pings. My worst hammer blows produce a dull, hesitant swipe, lacking the depth of a more committed 'thunk'. Silence can also be a sign of skill. If hammering isn't interspersed with loud swearing and the clanging of tools thrown in frustration, you could be dealing with an expert.

Stone in hand, filled with dread, I do everything right. I place it on a flat, stable surface. I support it in all the ways it needs to be supported. I measure and I mark and I pray. I raise my hammer with only the purest of intentions. I will not commit a murder.

Shit.

I curse the universe, my hands. I am a serial killer worthy of the longest podcast.

While Cal, a waller for more than twenty years, nonchalantly holds a stone in one hand and a hammer in the other, executing precise blows. His stone goes on to live a long and healthy life. Three children. A picket fence. Fucker.

Although the learning curve has been steep, after braving the tool bag, I am less likely to be found sitting amongst a half-ton of discarded stones, mute from exasperation. Shaping stone helps me to build more quickly and efficiently. Finding the right stone is harder (and slower) than making any stone the right one.

•

I smoothed out the folds of the information leaflet, scanning the list of possible side effects. Palpitations: *not too bad*. Psychosis: *far from ideal*. Skin peeling off: *Jesus Christ*. Sudden death . . . I reread, and it did indeed say 'sudden death', somehow managing to be both totally vague and completely definitive at the same time.

For some, methylphenidate is a miracle drug. For others, a tailspin that is almost impossible to exit. After years of executive dysfunction, the promise of basic function was tantalising. But compared to where I had been, I was doing well and I couldn't jeopardise that. It was greedy to want more than the best I'd ever had.

The thing was, even with Bupropion (the best I'd ever had), my potential had been dispersed all over my body in erratic, fast-moving particles. Occasionally, my brain would be lucky enough to catch one of them, and I'd glimpse what I was capable of.

The rest of the time was frustration and clinging to the hope and nostalgia of what had been, what could perhaps be again.

Tipping one of the elliptical white pills into my palm, I felt the florescence of hope. I reminded myself that I'd been here before, medication and expectation in hand. Swallowing the tablet, I tried not to let myself believe that it could do what hundreds before it could not.

Methylphenidate works the day you take it. The hour you take it.

Suddenly, there was a magnet inside me, attracting all the floating particles of potential into one place. They slowed to a stop and congregated somewhere near my prefrontal cortex. My thoughts began to play one track at a time.

I'd always lived intensely. Efforts confined to the day, the hour, the minute when motivation would strike. Never sure the feeling would last, or when it would return, I had to act immediately. Cleaning the whole house, writing the whole essay, moving the fridge, buying the rowing machine, ending the friendship. One frantic morning followed by three painfully lethargic, guilt- and anxiety-ridden weeks. Methylphenidate turned the rare occasion into the regular occurrence, so I no longer had to do it all at once. My panic lessened considerably.

I spent the first weeks in vacation mode, enjoying a new slower pace of life on a drug someone had once described to me, insanely, as 'speed'. As the medicated months passed, and I was able to do what I needed to when I needed to, it dawned on me that I had been locked in a tower, watching helplessly as the world went on without me. Methylphenidate was the key that unlocked the keep. The little white pill was a ticket to the ball.

And I'd arrived on time. The scratchy seams of my clothes didn't make me want to die. I ate only an appropriate amount of canapés. The music wasn't too loud. I remembered everyone's name, made small talk and calmly rejected the muscular prince with narcissistic tendencies. I was enjoying the dancing, people twirling around me, almost forgetting how I'd got there. Then, suddenly, the clock struck midnight. Even though I so desperately wanted to stay.

The post-5 p.m. stimulant crash was a reminder that it was only ever chemical sorcery. Returned to rags, irritable, hungry, I was again at the mercy of a brain beyond my control. Unable to make the kids dinner or open my email. But as months passed, evenings improved. The horse-drawn carriage was fine as a pumpkin. I just had to make sure I ate something and went to bed early.

Every morning, I lift my hefty and comprehensive pill box from its position on my nightstand, open one of the little plastic doors, and accept the help that falls into my hand. With chemical intervention, life is now liveable. And I revel in such an easy fix.

It took me two decades, thousands of hours of research and thousands of pounds of my own and my mum's money to find this meticulously balanced chemical stack I should really never refer to as my easy fix. Fix? Better, for sure. Easy? Not so much.

•

No one in the world but me could know there was tension in my mother's voice. It wound its way tight and subtly strained up the stairs to my bedroom. She was chatting with Janek, who was captive in our post-flood newly renovated kitchen, and she'd shown up hours later than I'd expected.

Aside from a brief mention of which day, she always skirted around the issue of when she was going to arrive. It was a point of contention: I wanted to know about when she would get here, and she did not want me to know about when she would get here. Holding my breath, I attuned my ears to her tone.

I already knew how this was going to go. The last carpeted step before the kitchen's cold tiles came too quickly.

'Decided to make an appearance?' She sat behind the steam

of a fresh cup of tea. Out of the corner of my eye, I saw Janek make a quick exit.

I busied myself with cups and tea bags and offered her a top-up.

'Actually,' she said, 'I'd like something to eat. I haven't had any dinner.'

It was 9.30 p.m. Dinner was hours behind us. I opened the fridge and, with a sense of foreboding, saw that we only had basics. I listed them, and she duly offered her disapproval, finally cutting me off with, 'I'll have to have a sandwich then.'

Minutes later, I watched her disassemble the cheese sandwich I had given her. She checked each slice of bread, raising her eyebrows to register a complaint about my stinginess with the butter.

'How's Nell?'

'She's great. Potato waffles and My Little Pony. Living the good life.'

'I brought the puppet theatre for her.'

I laid on thick enthusiasm. 'Oh, she'll love that. I won't tell her and you can surprise her when you get up tomorrow.'

'And Jo?'

'Struggling, but better in some ways. Ups and downs.'

Mouth full, she replied. 'Jo really should be getting out once in a while . . .'

I physically bit my tongue, pushing front teeth into meaty muscle. I'd explained this a thousand times.

'. . . it's not good for her to be in bed so much.'

How hard would I have to bite to draw blood?

Keeping my mum happy when she was determined to have a fight was like running through a constantly shifting minefield. Disagreement was fuel flung on the fire but agreement emboldened her.

'Jo being in bed is not a choice.' Despite my efforts, I sounded irritated. Her sandwich stopped at an indignant angle beside her mouth.

'I was just asking, Kristie.' she paused to let her offence land. 'She has a lot of the same issues that you did . . .'

Yes, my *issues*. My insurmountable issues. As she spoke, I found myself watching from outside in a way I had never managed before. I saw us. Me, my mum, at this messy kitchen table. Both so hurt and so determined. Two war-weary soldiers and a game of Russian roulette. The rules were simple and I knew them by heart: winner watched the loser self-mutilate.

'Don't worry,' I said. 'We're on top of it.' Not disagreement, not agreement. Commending myself on my neutrality and hoping, foolishly, to make a break for it, I started to load the dishwasher. My back was turned when I heard her, maudlin but incisive, 'You know, you were Jo's age when you changed.'

My breath caught in my chest. This old story. How she'd had a better Kristie but that one disappeared and she'd got stuck with me instead. One of her biggest hits, a classic. And it was working. I could feel my heart beating faster, harder. And this time my anger wasn't for me, but for Jo, unwittingly drawn into her nana's pathological need to pathologise. I concentrated on the pain in my hands, rinsing the cutlery under too hot water.

'You don't agree, but I wanted the best for you.' Here she sighed, as if it pained her to say it. 'There was something wrong with you and no one else was going to figure it out.' She took another bite before continuing casually, 'You would do the same for your child.'

She had loaded the bullet, spun the cylinder and set the gun on the table.

Do the same for my child? Did she mean *to* my child? Anything my mum had disliked about me had become part of my 'illness'. She had labelled the fragments that challenged her, the messy steps I'd taken towards becoming my own person: aberrant, diagnosable. She'd thought of me, and made sure I'd thought of myself, as a difficult person. No, a broken person.

It was what she'd presented to the outside world, too. Every doctor, every alternative therapist had been told (and accepted) that there was something wrong at home, and it sure as hell wasn't her. I had suffered because she couldn't see, or admit, the truth. But I saw so clearly now.

Turning around, I watched in amazement as I, without anger, stated a fact I knew in my heart to be absolutely true. 'No, I wouldn't. I choose every day not to do the same for my child.'

She looked at me, mouth slightly open. As if I'd spat on her.

A split second of absurd calm.

And then: 'You always find fault. So selfish. UNGRATEFUL. And I'm not the only one who thinks that. Your brother does, too. And your aunt. Don't get upset. I WON'T PROTECT YOU FROM WHAT YOU NEED TO HEAR.'

I moved away and sat on the stairs. A very angry person threw her phone and glasses into her bag, spilled the mug into the sink, demanded to know where her suitcase was. Gritted teeth. More shouting. I WAS CROSSING HER BOUNDARIES. I said nothing.

She picked up her coat by its fur-lined collar and looked straight at me. She spoke slowly. 'Are you still in therapy, Kristie?'

I saw myself, hunched against the wooden banister, dishcloth in hand.

Her voice was sweet but her face was alight, willing me to engage. 'Because you need to be.'

Grenade in hand, fingers looped through its metal pin, I waited for pain to kick in the door. She was still talking, saying that I was delusional, so fucked up I needed a care worker. I would burn whatever was left to the fucking ground.

Yet beneath the fizzing static of my emotional edges, there was a core stillness. I remembered how I had felt amidst the havoc of my childhood but the pain stayed in the past. I hadn't been able to look after myself back then. I could now.

And I heard my own voice, measured, telling her that if she didn't stop she was going to have to leave.

'I won't stay where I'm not welcome,' she was saying. 'I'M NOT BEING FUCKING TREATED LIKE THIS.'

My first movement was an involuntary flinch when the door slammed and rattled in its frame. I could hear her outside, ranting through tears. I stayed on the stairs until her car pulled jerkily out of the drive and Janek, emerging from the shadows, asked what had happened, how I was. I could only shrug.

I thought about my mother's pain. How it had been bestowed upon her, how she had forced it upon me. How young I had been when it found a place in my life and my thinking. A vast celestial body that drew other pain into its orbit, taking up parts of me meant for me. For so long that had been my whole world.

But now it wasn't.

Here, in this moment, I felt untangled. A thread all my own.

I reached for my phone, dialled her number. It rang only once. She answered. Silence.

'Come back, Mum. It's late, you're upset, and you shouldn't be driving.'

'OK,' she said, her voice breaking.

I waited a few seconds before hanging up.

Back on the stairs, I sat quietly, listening for the sound of tyres on gravel. When she pulled in I got up, opened the door and peered out into the dark. She came towards me, eyes downcast, suitcase dragging.

Hugging her with leaden arms, the words came lightly, clearly. I told her I loved her, I was glad she had come back, but I was going to bed.

I could feel her shoulders heaving. 'I'm sorry, Kristie.'

A while later, I checked on her. I watched my mother sleep, dark hair in a grey-flecked halo on the pillow, body small in a bed that wasn't hers. I breathed quietly and carefully through a growing strain in my chest, threatening to be cracked open by the burgeoning pain of understanding.

Lifting the Copes

Arching my back, I push out my belly and rest the weight of the stone in the space between my ribs and my navel. This is extremely shitty lifting technique. I loosen my fingers from their tight grip and let the stone fall to the ground. Trying again, I keep my back straight and, arms shaking, quickly weakening, I heave the cope stone onto the wall. Proportionally, cope stones look small but they are this wall's largest.

The icing on the drystone cake, copes are decorative, and can be arranged in many styles. (Upright, double, flat, stag and doe.) They also perform a structural role: holding the wall together at its highest point and stopping outside elements from making their way in.

I settle the cope on the wall, making sure it's straight before stacking another beside it. Another follows. This last one needs some shaping, and I begrudgingly lift it back off the wall and onto the banker. I'm tired, body aching, but so close to finishing. And so goes the lifting, settling, straightening, lifting, shaping, lifting, settling until the whole length of the wall is topped, a tidy bookshelf of vertical stone. I step back, and sitting on the upturned wheelbarrow I take off my gloves as I look over my work.

Smiling to myself, I think, I built that.

Copes complete the transformation of a pile of rough stone to a standalone structure that can resist the ravages of weather and time. Never perfect, never finished, but ready to perform its function as a boundary, a shelter, a place for things to grow.

I built that.

It looked like a medieval torture device as my mum stretched coiled springs to hook curved metal over the edges of blue and white china. She passed me the plate and tapped a nail into the wall with a hammer longer than her forearm. I looked at Rabbie Burns' face printed on the porcelain. His high collar and wispy sideburns, surrounded by the last verse of 'To a Louse'. He stared at me from that wall for years.

When I did something my mum didn't like, she would quote the words from the plate. 'O wad some Power the giftie gie us, To see oursels as ithers see us!' Weaponisation of the standard Habbie. To ensure that I would heed her opinion, and to convince me it was a gift, the overpriced kind for which I should be especially grateful. The world taught me the same lesson in less lyrical ways, and I learned to measure my worth by my (generally underwhelming) capacity to be liked. If I neglected to know people and their opinions, I might never know myself. After all, who was I to say who I was?

As careers go, drystone is off the beaten track, literally. There are early starts on back roads to places that, according to Google Maps, don't exist. Days where we don't see a single other person. Hours pass, marked only by hammer blows. 'Where's the wee chisel?' 'Tea?' 'Ugh.'

Me, Kristie De Garis, a woman of few words.

On the drive home, I ponder the work I've done and the work I have yet to do. Shower, dinner, quiet time with the kids before bed, finally giving in to sleep. Where once I fervently sought the company of others, my life has shifted, adjacent to the rest of the world.

I'd always been right in it. Welcoming noisy distraction after noisy distraction, living my mum's interpretation (and probably not the spirit) of Burns' poem. At first, all I could hear in this new seclusion was my own mad ruminating. But soon, mind and body steadily occupied by drystone, there was no room for that. What remained was a lot of clarity without much thinking.

Me, Kristie De Garis, a woman of few thoughts.

And the clarity stays with you. It turns out there is a significant difference between who I was in the busy world and who I am at my core. Here, stone and tools in hand, the smell of metal and sweat mingling on warm skin, I don't have to be anything but myself: a hyper-discerning rural introvert.

Scraping away the opinions of others takes time. Hands dirty, I am unearthing my long-buried self from the ground. The pieces of a person who can exist wholly and separately from the world.

•

Both imagination and a fondness for procedure are required to successfully build drystone. After all, it is a craft that creates strict order from the chaos of a stone pile. Although wallers each have a way they prefer to work, their approaches are built around the same core techniques. Some separate stone before they start building, some build directly from the stone pile, some prefer to use power tools, some see that as sacrilege, but a cheekend is always built the way a cheekend is built.

I'd learned the 'rules' of drystone as:

Stones length in.

Hearting placed tightly.

Cross your joins.

Build with the batter of the wall.

Keep your stones level.

Though these five simple principles largely kept me right, I have also learned how easy it is to lose sight of the big picture.

Once you understand the basics, there's not a lot else to do but put stones in the wall. A natural state of hyperfocus. There are days when I won't walk away from the wall until it is time to go home. And that's when I notice the mistakes. Taking down sections of wall I have painstakingly built is soul-destroying and so I have added another rule to the list.

Step back.

Step back and walk away. Give yourself space to regard your work as a whole, rather than solely in relation to the last stone you placed.

Admire your wall. Trace it with your eyes, enjoy the lines. Find that difficult stone, now sitting in place, remember the struggle and know you have done your work. In looking, you may also notice a course looking less than level, a stone protruding too far, the beginning of a running joint or a zipper, or perhaps just a section that looks clumsily built.

By regularly stepping back, oversights and issues remain manageable, don't develop into unsolvable problems down the line. Step back, and you won't find yourself building on top of mistakes.

But no matter what you see, it's important to remember that there was a time, so many times, when even this was beyond you.

Afterword

I live a carefully curated life. Some might describe me as a hermit, but I only live in solitude if you count seeing very few other people as solitary. My days are filled with connection.

One of my greatest accomplishments is being able to offer my children the support they need to become themselves in a world that actively discourages it. I am honoured every day to know them as they really are, and for them to know me in the same way. When we talk, we don't first need to navigate a barricade of wounds and pain. The love we feel for one another isn't slowed by complex defences; it moves easily and directly, heart to heart.

And then there's my relationship with the natural world. The flowers, the hills, the land, the light and the stone.

If you ever find yourself at Lunan Bay in Angus, head towards the cliffs at the east end of the beach. If the tide is out, you can walk a tight maze of sandy channels between huge rocks. The colours, textures and lines are mesmerising, and I have spent hours there photographing details created by the coming together of sea and stone.

My mum has often told me she has a strong connection with trees. My strongest affinity is with stone. In a world I find overwhelming, it quietly absorbs the chaos. Standing at Lunan Bay among the sea-smoothed monoliths, hand against their sun-warmed surfaces, feet sinking in the still wet sand, the stone tunes out the unnecessary. I'm grounded, aware, totally present. My body even feels different, more neutral. This is my church.

Acknowledgements

I am fortunate to say that I have many people to thank. Not just those who helped and supported me in the very specific task of writing this book, but every person who got me here. Even if we are no longer in each other's lives, if you treated me kindly when I could not do that for myself, thank you.

A, P, L and J – thank you for making all the space for this.

To my mum and my brother, it's been so hard for all of us. Thank you for always doing your best.

Svea, you are an alchemist. You coaxed a writer (and a book) out of me. Thank you for everything including but not limited to: your skill, time, generosity, empathy and honesty. The poo really does block the pee.

Writing this book is one of the hardest things I've ever done. Of course joyous, cathartic and compelling, but also intense, painful and boring. Many people helped me many times to pull through. Gemma, Kara, Conor, Alec, Callum, Kate, Jesse, Eileen, Rodrigo, Lili, Hanna, Andy, Alison and Jason, thank you for your eyes and your ears, your friendship, enthusiasm and encouragement.

Thanks to Jamie Crawford for seeing this book before I did. To Alison Rae for her skill, clarity, understanding and patience (I sent a lot of emails, asked a lot of questions). And to the rest of the team at Birlinn who pulled it all together. It takes many people to make a book. I am grateful to have been so supported through this process.

Thank you to Cristy and the rest of the team at Tomnah'a for allowing me to spend so much time with the flowers, and for all the hard work that goes into making the garden such a beautiful and healing space. Thank you to members of the COVID Informed Community (both online and 'in real life') for helping us navigate Long COVID, and this ongoing pandemic, with our sanity relatively intact.

To therapy, to medication, to sobriety, to diagnosis, to the land, to the stone and to my small awkward self with sturdy legs and her hair sticking out on one side. We've worked so fucking hard.